Brain Neurosecretory Cytokines

Immune Response and Neuronal Survival

Brain Neurosecretory Cytokines

Immune Response and Neuronal Survival

Armen A. Galoyan

Institute of Biochemistry
Yerevan, Armenia

Kluwer Academic / Plenum Publishers
New York, Boston, Dordrecht, London, Moscow

Library of Congress Cataloging-in-Publication Data

Brain neurosecretory cytokines : immune response and neuronal survival / edited by Armen
A. Galoyan.
p. ; cm.
Includes bibliographical references and index.
ISBN 0-306-48337-8
1. Neuropeptides--Physiological effect. 2. Cytokines--Physiological effect. 3.
Neurohormones--Physiological effect. 4. Neurons. I. Galoyan, Armen, 1929-
[DNLM: 1. Cytokines--physiology. 2. Neurosecretion--physiology. 3.
Neuroimmunomodulation. QW 568 B814 2004]
QP552.N39B725 2004
612.8'22--dc22

2003070341

ISBN 0-306-48347-8

233 Spring Street, New York, New York 10013

http://www.wkap.nl/

10 9 8 7 6 5 4 3 2 1

A C.I.P. record for this book is available from the Library of Congress

Printed in the United States of America

CONTENTS

INTRODUCTION

Discovery of neuroendocrine cells of hypothalamus nuclei, *N. Paraventricularis* (NPV) and *N. Supraopticus* (NSO)[1], elucidation of vasopressin and oxytocin primary structure[2] in the fifties as well as isolation and chemical identification of releasing hormones[3,4] can be considered as the foundation of modern molecular neuroendocrinology and neurochemistry of biological active peptides and proteins of hypothalamus-neurohypophysial system. Moreover, synthetic analogs of releasing and inhibitory neurohormones (LH-RH, GH-RH, somatostatin, and bombesin/GRP) are effective drugs for treatment of various tumors[5]. During 40 years we added new dimensions to this area. Simultaneously with investigations of hypothalamus releasing hormones by American scientists we made our own contribution to this area:

1) Discovery of the functional system "neurosecretory hypothalamus-endocrine heart" which is important not only as a system itself but for the adaptive capacity of the organism in general. Isolation of cardioactive hypothalamic proteins representing specific carriers and precursors of coronary dilating neurohormones K, C, and G[6,7] can be considered as an important event in this field. Cardioactive protein-hormonal complexes (PHC) are localized primarily in the synaptosomes as well as in the neurosecretory granules of hypothalamus-neurohypophyseal system. In various cardiac disorders, like ischemia and myocardial infarction, sharp reduction in PHC occur in different regions of the brain parallel with its multiple elevations in several organs and tissues including several-fold increase in the blood.

In 1967 we discovered the phenomenon of peptide hormone biosynthesis in atrial ganglionic cells and peptidergic fibers of atria[8]. There is neurohumoral feed-back between NPV- and NSO-produced coronary dilatory and coronary constrictory peptide hormones and atrial peptide hormones, which are releasing factors for hypothalamic cardioactive hormones K, C, and G in general circulation. This new class of hypothalamic cardioactive neurohomones in conjugation with hormone production function of ganglionic atrial cells allows us to approach closely to the understanding of new principles of interrelation between cardiac endocrine system and hypothalamic neurosecretory hormone production purposed for maintenance of optimal conditions for brain and heart functioning, as well as for adaptation of the organism as a whole[9-32].

2) Discovery of the brain neuroendocrine immune system. Considering immunomodulators of the brain immune system (hypothalamic neurosecretion of cytokines) as an inalienable part of the adaptation mechanism, in the last period our investigations (1986-2000), we concentrated our search on brain signal molecules of immune system produced by neurosecretory cells of NPV and NSO and on the role of these substances in providing (in cooperation with glial cells) the protection of the brain during various infections and degenerative disorders and in the regulation of immunocompetent cells in the organism[31-37].

Our data on the biosynthesis of a number of known interleukins and novel cytokines in neurosecretory neurons of hypothalamus (NPV and NSO) demonstrate that those neuroendocrine nuclei of the hypothalamus are centers of neuroendocrine-immune system of the brain. Results of our investigations on the biology of hypothalamic neurosecretory cardioactive hormones and immunomodulators were summarized in my two monographs «Some problems of biochemistry of hypothalamic regulation» (1965)[7] and «Biochemistry of novel cardioactive hormones and immunomodulators of the functional system neurosecretory hypothalamus - endocrine heart» (1997)[31] as well as in more than 500 articles. We have succeeded in the discovery, purification from neurosecretory granules of bovine hypothalamus and neurohypophysis, and determination of the primary structure of a series of novel cytokines (interleukins) and other signal molecules of brain neuroendocrine immune system (Table 1). Use of these new classes of hypothalamic neurosecretory peptides should lead to a more effective treatment of immune, neurodegenerative, and neuroendocrine diseases. As was shown, we demonstrated also some of the main biochemical mechanisms of their action.

According to immunohistochemical data, the biosynthesis of interleukins, immunophilin - receptor of FK-506 immunosuppressor, and thymosin β_1 take place in NSO and NPV. In 1992 we discovered in bovine hypothalamus immunophilin (M_r 11778) which possesses peptidyl-prolyl-cis-trans-isomerase activity and is a receptor for immunosuppressant FK-506. It was the first work on immunophilin isolation from the brain, and we had established the complete primary structure of the immunophilin. With antibodies to its N-terminal fragment (1-15) we showed that this protein is a localized in the neurons of NSO and NPV, both in rat and frog hypothalamus. These data provided evidence of important role of immunophilin in the neuroendocrine cells of hypothalamus, probably the participation in the biosynthesis of interleukins in these cells. As it is known, calcineurin is a target for FK-506-FKBP (immunophilin) complex in the cells. Mainly via this mechanism, the transcription of interleukin-2 gene is inhibited in lymphocytes, which is accompanied by suppression of immune response. It can be suggested that similar complexes may be formed between immunophilin and its endogenous ligands (we showed that cardioactive neurohormone C, $T\beta_4$ (1-39), and several other hypothalamic peptides exhibited sufficiently strong binding to hypothalamic immunophilin), hence, immunophilin can participate in the regulation of interleukin biosynthesis in the neurosecretory cells of hypothalamus.

We isolated from bovine hypothalamus macrophage migration inhibitory factor (MIF), which originally was described as one of the first cytokines secreted by activated lymphocytes and capable of inhibition of macrophage random migration in vitro. More recently MIF was characterized as an immune mediator counteracting the action of glucocorticoids. Precise physiological functions of MIF-1, MIF-2, and MIF-3 in relation to the immune response in CNS remain is to be elucidated.

Table 1. Discovery of the hypothalamic pool of immunomodulators: primary structure and functions (Galoyan et al., 1986-2000)[31-37].

POLYPEPTIDES	PRIMARY STRUCTURE	THE MAIN BIOCHEMICAL MECHANISMS OF ACTION & FUNCTIONS
THYMOSIN β_4(1-39)	ACETYL-SDKPDMAEIEKFDSK-LKKTETQEKNPLPSKETIEQEKQ	ACTIVATOR OF Ca^{2+}, CaM-DEPENDENT ENZYMES (cAMP PDE, MLCK etc) WITHOUT PARTICIPATION OF Ca^{2+} & CALMODULIN
THYMOSIN β_4(1-4)	ACETYL-SDKP	INHIBITOR OF cAMP PDE & MLCK
MACROPHAGE MIGRATION INHIBITORY FACTOR ISOFORMS (MIF-1, MIF-2 & MIF-3) (12 369.2, 12299.7, 9496.2 Da)	N-TERMINAL SEQUENCE: PMFVVNTNVPRASVPDGLLSE-LTQQLAQATGKPAQYIAVHVVP	ROLE OF BRAIN MIF ISOFORMS IN THE REGULATION OF A VARIETY OF FUNDAMENTAL NEUROIMMUNOLOGICAL PROCESSES IS UNDER STUDY
IMMUNOMODULATORS ISOLATED FROM NEUROSECRETORY GRANULES		
INTERLEUKINS: IL-1α, IL-1β, IL-2, IL-6, TNF -α		
IMMUNOPHYLINS (FK-506 BINDING PROTEIN) Mr 11778.4; 11825.5	GVQVETISPGDGRTFPKRGQTC-VVHYTGMLEDGKKFDSSRDRN-KPFKFMLGKQEVIRGWEEGVA-QMSVGORAKLTISPDYAYGA-TGHPGIIPPHATLVFDVELLKLE	PEPTIDYL-PROLYL-CIS-TRANS-ISOMERASE, RECEPTOR OF FK-506 IMMUNOSUPRESSOR. FK-506 PROTECT OF EMBRIO HIPPOCAMP Al & CORTICAL NEURONS NEUROTOXICITY (*IN VITRO*) INDUCED BY GLUTAMATE & BETA-AMYLOID (25-35), INCREASE ADRENALINE & NORADRENALINE RELEASE FROM ADRENAL CHROMAFIN CELLS *IN VITRO*. *N*-TERMINAL (1-15) IMMUNOPHILIN (1μm) RESULTED IN A PRONOUNCED INCREASE OF THE ENDOGENOUS GLUTAMATE RELEASE BY BRAIN CORTEX SYNAPTOSOMES.
THYMOSIN β_1 (UBIQUITIN)	MQIFVKTLTGKTITLEVEPSDT-IEDVKAKIQDKEGIPPDQQ-RLIFAGKQLEDGRTLSDYNI-QKESTLHLVLRLR$_{74}$ – GG$_{76}$	CYTOPROTECTIVE, SMALLEST H S P_S, CALMODULIN ANTOGONIST IN THE PRESENCE OF Ca^{2+} (0,2mM) *IN VITRO*
NEW CYTOKINES		
PROLINE-RICH PEPTIDES	AGAPEPAEPAQPGVY AGAPEPAEPAQPGV AGAPEPAEPAQPG APEPAEPAQP	REGULATORS OF LYMPHOID AND MYELOID CELL LINES, NEUROPROTECTORS, ANTIBACTERIAL DRUGS, (*IN VIVO*), INHIBITORS OF JURKAT CELLS PROLIFERATION *IN VITRO*

In 1992 we succeeded for the first time in isolating from bovine hypothalamus of a protein which in concentrations 10^{-9} - 10^{-12} M activated Ca^{2+}-calmodulin-dependent enzymes (cAMP PDE, MLCK, etc.) without participation of calcium ions and calmodulin. After separation of individual components of this protein preparation, the complete primary structure of one of these proteins was established by microsequencing and mass-spectral analysis[37]. This protein was thymosin β_4 (1-39). Tβ_4 (1-39) was shown to be the primary regulator of Ca^{2+} - calmodulin-activated enzymes, since for the activation of calmodulin, Ca^{2+} was necessary in a concentration of 10^{-5} M. Thymosin β_1 (1-74 and 1-76) (ubiquitin) as well as a new group of peptide hormones (proline-rich peptides) were isolated from the neurosecretory granules of bovine neurohypophysis. These compounds were shown to be powerful regulators of the immune system[32]. The role of thymosin family (Tβ_4, Tβ_9, Tβ_{10}, etc.) in the immune system is well characterized.

Data obtained prove the presence of numerous immune molecules in the neurosecretory granules, which in its turn provides evidence for the existence of neuroendocrine immune system in the brain. Important is to mention that the brain requires immunological defense during neurodegenerative diseases, especially if pathological process is aggravated with infection. The problems of autoimmune and neurodegenerative diseases of brain (Alzheimer's and Parkinson's diseases) are the subject of neuroimmunological investigations. To autoimmune diseases of brain belong encephalomyelitis, spongiform encephalomyelopathy, paraneoplastic encephalomyelitis, autoimmune neurological disorders, autoimmune disease induced by intracellular vesicle-associated proteins, multiple sclerosis. Several neurodegenerative diseases of the brain are characterized by elements of autoimmune protein formation like it takes place tumor cells. Discovery of a large group of brain immune system signal molecules synthesized by NSO and NPV allow me to suggest that these molecules play a role in the mechanisms of immune response and neuroprotection during viral and bacterial infections, autoimmune and neurodegerative disorders, trauma, etc.

Viral infection can be transported into the brain parenchima through the axons (retrograde pathway)[38]; peripherally injected lipopolysaccarides (LPS) induce biosynthesis of IL-1 in the hypothalamus. Subdiophragmal vagotomy effect is blocked[40].

Immunological challenge alters electrical activity in the hypothalamus[41] and activates the hypothalamic-pituitary-adrenal axis (HPAA)[42,43]. Through the afferent fibers of n. vagus the chemical information is transported to n. tractus solitarius (NTS) and directly to NPV[44]. The brain can defend itself with its reserve immune bodies when the blood-brain barrier (BBB) is not damaged but virus infection can be transported into the brain.

Present monograph summarizes our data concerning chemistry and biology of signal molecules of brain neuroendocrine immune system, mainly proline-rich peptides, playing important role in the regulation of immune response, neuronal survival, hematopoiesis, and adaptation as a whole.

1

NEUROSECRETION OF CYTOKINES BY N. PARAVENTRICULARIS AND N. SUPRAOPTICUS OF HYPOTHALAMUS

1.1. THE ISOLATION OF NEUROSECRETORY GRANULES FROM BOVINE HYPOTHALAMUS AND NEUROHYPOPHYSIS: BIOASSAY METHODS OF CYTOKINE IDENTIFICATION

Secretory granules were isolated from bovine hypothalamus and neurohypophysis, dissected out immediately after the animals were slaughtered. Homogenization of the tissues and the procedures of granule preparation, purification, and lysis were carried out in the medium containing protease inhibitors (phenylmethylsulphonyl fluoride, benzamidine, and leupeptin)[31,45].

Experiments directed at neuronal cytokine-like activity determination were performed with lysates of the neurosecretory granules.

Protein concentration in the granule lysates was determined according to Ohnishi and Barr using BSA as a standard[46].

The presence of IL-1α and β activities in the granule lysates was assayed by their ability to maintain the proliferation of the T-helper cell clone D10.G4.1[47-49]. The cells (2 x 10^4) in 200 μl of a culture medium containing 2.5 μg of concanavalin A (ConA) were incubated for 72 h at 37°C in the presence of 7% CO_2 with various concentrations of the granule lysates. During the last 6 h 0.5 μCi of [^{3}H]-thymidine was present in the medium. A quantity of IL-1 which induced 50% of maximal proliferation of cells was accepted as one unit of IL-1 activity. The inhibition of IL-1α and β activities was achieved by the addition of rabbit antisera to these ILs.

The presence of IL-2 activity in the granule lysates was assayed by their ability to maintain the proliferation of the IL-2-dependent line of the cytotoxic T-lymphocytes (CTLL-2)[50,51]. The cells (2 x 10^4) in 100 μl of the culture medium were incubated for 24 h at 37° C with various concentrations of the granule lysates, individual neuropeptides, or standard IL-2. Six hours before the end of the incubation 0.5 μCi of [^{3}H]-thymidine was added. Neutralization of IL-4, which can also maintain the proliferation of the CTLL-2 cells, was achieved by the addition of a monoclonal antibody to IL-4 (11B11). The results were evaluated by comparison with the activity of IL-2 standard (100 U/ml).

IL-6 activity in the granule lysates was assayed by their ability to maintain the growth of the IL-6-dependent mouse B cell hybridoma (line B 9)[52]. The cells (2.5×10^3) in 200 ml of the culture medium were incubated for 72 h at 37°C with various concentrations of the lysates of neurosecretory granules of neurohypophysis (NGN). During the last 6 h 0.5 μCi of [^{3}H]-thymidine was present in the medium. A quantity of IL-6 which induced 50% of maximal cell proliferation was accepted as one unit of IL-6 activity.

Cytotoxic activity of TNF-α in the granule lysates was assayed by its ability to lyse the fibroblast cell line L_{929} (sensitive to TNF-α)[53,54].

The TNF-α neutralization was conducted by the use of rabbit antiserum to mouse TNF-α. The protein quantity which induced 50% lysis of fibroblasts was accepted as one cytotoxic unit. It was of great interest to detect TNF-α in the neurosecretory granules taking into account that TNF-α produced in the brain[55] can block the stimulation of the HPA axis that occurs during neuronal inflammation.

The ability of NGH (neurosecretory granules of hypothalamus) and NGN lysates to affect the expression of inflammatory cytokines IL-1 α and β, IL-2, IL-6, and the secretion of TNF-α was studied on astrocytes obtained from a primary glial cells culture of BALB/c mice[56]. Astrocytes (2×10^4) were kept in the culture medium for 8 h, then the neurosecretory granule extracts, LPS, and polypeptides were added, and the incubation continued for 24 h. The interleukins were determined in the supernatants of the LPS-stimulated glial cells by the above mentioned methods.

Antigen presentation was assessed by measuring the ConA–specific T helper cells clone D10.G4.1 proliferation in response to ConA–pulsed astrocytes or peritoneal macrophages[57].

1.2. IDENTIFICATION OF INTERLEUKINS 1α AND 1β IN THE LYSATES OF NEUROSECRETORY GRANULES OF BOVINE NEUROHYPOPHYSIS AND HYPOTHALAMUS

The effects of the NGH and NGN lysates on the proliferation the IL-1 (α and β)-dependent T-cells clone D10.G4.1 are presented in Figs. 1 and 2, respectively. In the presence of both NGH and NGN lysates the IL-1 (α and β) activities were enhanced. Hence, the lysates of NGH and NGN themselves possessed IL-like activities. The activities in both cases were dependent on protein concentrations in the granule lysates. Thus, the NGH lysate in a dilution of 1:10 possessed the highest activity. After addition of rabbit anti-IL-1α immune serum to the cell culture the interleukin activities decreased by 75% (Fig. 1B). Both IL activities were inhibited entirely by simultaneous addition of antisera to IL-1α and IL-1 β (Fig. 1C). Relative activities of IL-1α and ILβ were approximately 75% and 24% (Fig. 1 A, B, and C). In the case of NGN the highest activity was observed at a dilution of the NGN lysate of 1:20 (Fig. 2A). The simultaneous addition of antisera to IL-1α and IL-1β to the culture removed completely the effect of the NGN lysate (Fig. 2C). The differences in IL activities shown in Fig. 2 A and B are 76% for IL-1α and 24% for IL-1β. The experiments demonstrated that the activities of both IL-1 α and β in NGN were higher than in NGH. The total activity of IL-1 α and β in control experiments was 15.6 U/ml; the addition of the NGH lysate (1:10) caused an increase in the combined activity to 65.5 U/ml, and the addition of the NGN lysate – to 84.3 U/ml. Taking into consideration that the initial protein concentration in the NGH lysate was 37.5 mg/ml and in the NGN lysate - 19.4 mg/ml, we can conclude that the IL1 concentration in the neurohypophysis was twice as high in comparison with that in the hypothalamus.

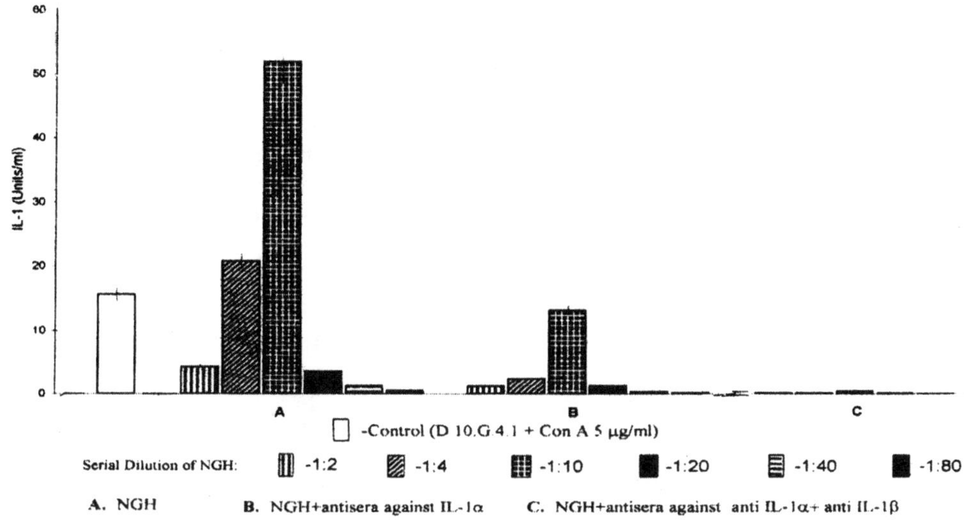

Figure 1. Presence of IL-1α and IL-1β activity in the lysates of neurosecretory granules of hypothalamus (NGH).

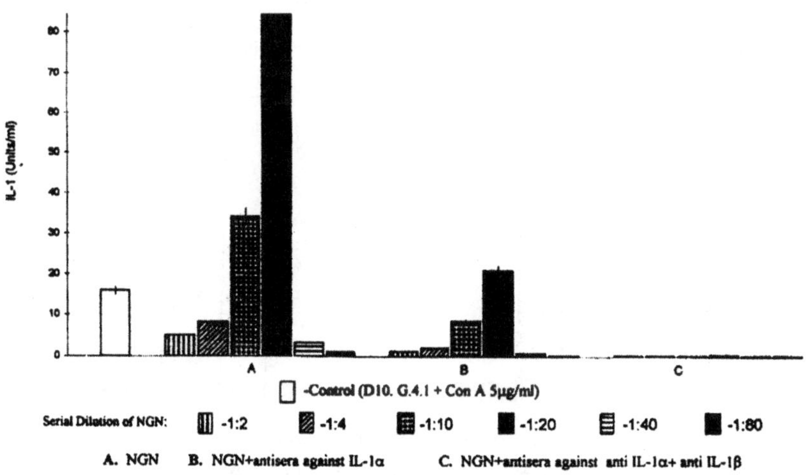

Figure 2. IL-1 activity in neurosecretory granules of neurohypophysis (NGN)

1.3. IDENTIFICATION OF IL-2 IN THE NEUROSECRETORY GRANULES OF BOVINE NEUROHYPOPHYSIS AND HYPOTHALAMUS

It was reported that the neuroendocrine hormones vasopressin and oxytocin are capable of replacing the IL-2 requirement for gamma-interferon production[58]. However, vasopressin and oxytocin do not induce incorporation of [^{3}H]-thymidine in DNA.

IL-2 activity was also revealed both in NGH and NGN lysates by their ability to stimulate the proliferation of the IL-2-dependent line of cytotoxic T-lymphocytes (Figs. 3 and 4). We used a standard IL-2 solution (100 U/ml) to maintain the proliferation of these cells in control. As follows from Fig. 3 A, IL-2 activities in the NGH lysate dilutions were 63.4, 6.1, and 0.57 U/ml at the following protein concentrations in the lysate aliquots added: 3.75 mg, 375 μg and 37.5 μg per 100 μl of culture medium. After addition of anti-IL-4 monoclonal antibodies to the culture, the activity of IL-2 remained unchanged (Fig. 3 B). As shown in Figs. 3 and 4, the IL-2 activity was higher in the NGN lysate in comparison with that in the NGH lysate (82.3 and 63.74 U/ml, respectively).

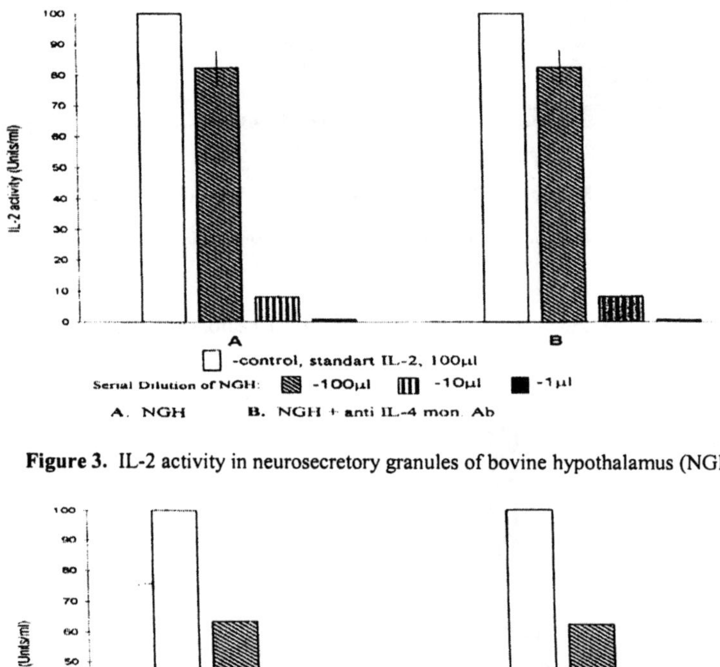

Figure 3. IL-2 activity in neurosecretory granules of bovine hypothalamus (NGH).

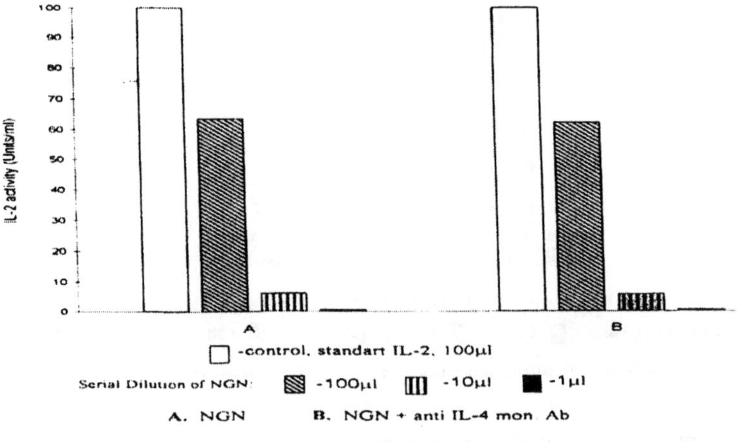

Figure 4. IL-2 activity in neurosecretory granules of bovine neurohypophysis (NGN).

Bearing in mind the fact that the protein concentrations per one ml of the NGN and NGH lysates were 19.4 and 37.5 mg/ml, respectively, it becomes obvious that the concentration of IL-2 in the neurohypophysis was two times higher than that in the hypothalamus.

1.4. IDENTIFICATION OF IL-6 IN THE NEUROSECRETORY GRANULES OF BOVINE NEUROHYPOPHYSIS

IL-6 is produced by a variety of cells, including macrophages, T cells, and B cells. We were able to observe the existence of IL-6 activity in the NGN lysates. The experiments performed with IL-6-dependent mouse B cell hybridoma showed that the activity of IL-6 with 100 μl of lysate was lower than that with 50 μl (Fig. 5). The level of activity did not depend on the time of the culture incubation (72 and 96 h).

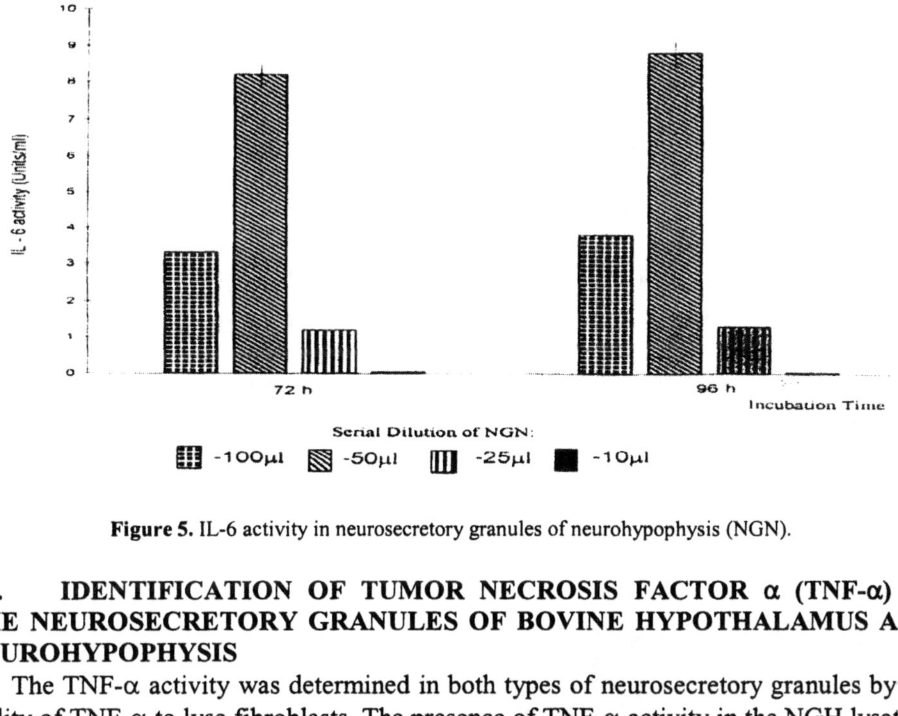

Figure 5. IL-6 activity in neurosecretory granules of neurohypophysis (NGN).

1.5. IDENTIFICATION OF TUMOR NECROSIS FACTOR α (TNF-α) IN THE NEUROSECRETORY GRANULES OF BOVINE HYPOTHALAMUS AND NEUROHYPOPHYSIS

The TNF-α activity was determined in both types of neurosecretory granules by the ability of TNF-α to lyse fibroblasts. The presence of TNF-α activity in the NGH lysate is illustrated by Fig. 6. TNF-α activities in the NGH lysate were 40, 18.2, 10.3, 1.4 U/ml at protein concentrations in the lysate aliquots of 3.75, 1.87, 0.94, and 0.37 mg/100 μl of culture medium, respectively. Monoclonal antibodies to TNF-α abolished completely the effect of granule lysate on fibroblasts (Fig. 6B). Analogous results were obtained with the NGN lysate (Figs. 7A and B). In the latter case the highest activity was obtained at the addition of 0.97 mg of protein per 100 μl of culture medium. In the case of the NGH lysate maximal TNF-α activity of 40 U/ml was observed with of 3.75 mg of protein per 100 μl of culture medium, and in the case of the NGN lysate the maximal activity of 60 U/ml was obtained with 0.97 mg of protein per 100 μl. These results let us conclude that

TNF-α content in NGN was 3 - 4 times higher than in NGH. Work on the role of neurosecretory ILs in brain functions and the regulation of hypothalamo-hypophyseal-adrenal axis is in progress.

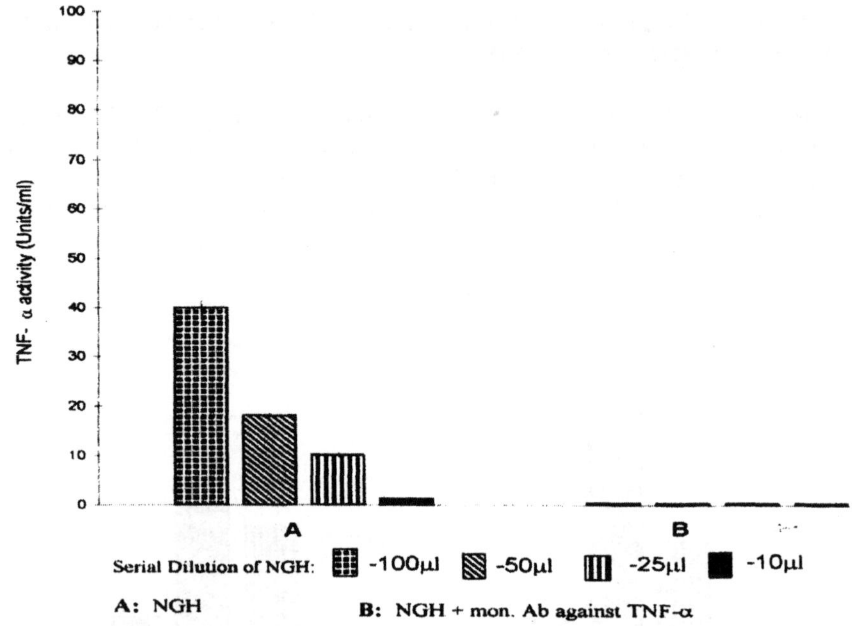

Figure 6. TNF-α activity in neurosecretory granules of hypothalamus (NGH)

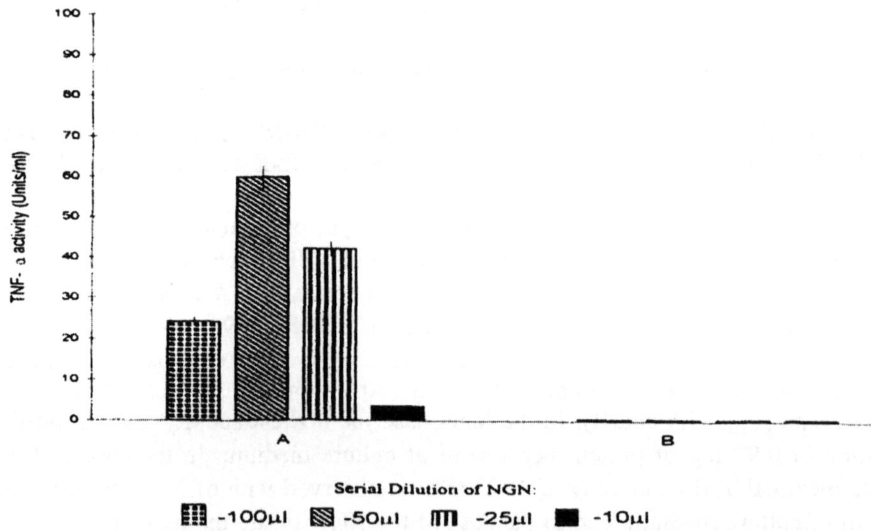

Figure 7. TNF-α activity in neurosecretory granules of bovine neurohypophysis (NGN).

The CNS can regulate specific immune system functions through the production of neurotransmitters, neuromodulators and neuropeptides, while the immune system can regulate specific functions of nervous system through the production of immunomodulators and immunopeptides[74-76]. A complex immune-neuroendocrine network operates under both physiological and pathological conditions[59-61].

We established that cytokines produced by neurosecretory cells of NPV and NSO arrive in the neurohypophysis by axonal transport and are released in the general circulation[31]. The capacity of cytokines to modulate the activity of the HPA axis was described by H. Besedovsky and coworkers in 1981[42,65]. They demonstrated that cytokines injected directly into the brain could induce neuroendocrine changes. Several immune-derived products, in particular several cytokines, can mediate neuroendocrine responses during immune responses[65]. "The results show that stimulation of peripheral immune cells induces cytokine expression in the brain, since peripheral IL-1 administration results in IL-1 gene expression in the CNS".

A large body of experimental evidence indicates that exogenous cytokines affect the metabolism, the neurophysiological and neuroendocrinological processes, mainly the HPA[42,43,65,66].

N. Rothwell writes[66]: "... It has been shown that IL-1 injected into the cerebral ventricles or the brain parenchyma, at doses as little as picomoles, induces responses which mimic those that occur during disease and injury, i.e. fever, anorexia, sickness behavior, slow-wave sleep, and alterations in neuroendocrine (e.g. activation of the hypothalamic pituitary adrenal axis), cardiovascular, and immune system function. IL-1 appears to act as an endogenous mediator of these diverse host-defense responses, mainly through actions in the hypothalamus, since localized injection of IL-1α or antibodies to IL-1β attenuate such responses to systemic stimuli. There is some evidence to suggest that IL-1β is the primary mediator of neurodegeneration. IL-1β is the predominant form of IL-1 induced by brain insults, and central administration of neutralizing IL-1β antibody inhibits damage caused by reversible cerebral ischemia in the rat".

My suggestion is that physiological significance of endogenous neurosecretory cytokines differs from that of exogenous injected cytokines.

Thus, although most cytokines are expressed in the healthy adult brain, the function and actions of many of these cytokines in the brain, particularly of cytokines produced by neurosecretory cells of hypothalamus, remain to be elucidated.

"It is worth noting that effects of cytokines when injected or induced either peripherally or centrally, are in many cases remarkably similar. In our opinion, these similarities indicate that the production of both peripheral and central cytokines underline well-programmed steps of responses integrated at brain level"[65].

Prof. H. Besedovsky's statement about the similar affect of peripherally and centrally administered cytokines seems not to leave any room for addition of endogenous interleukin effects produced by neurosecretory cells (NPV and NSO of hypothalamus), particularly on the hypothalamus-neurohypophisis system. However, the fact of interleukin biosynthesis in the neurosecretory cells of the hypothalamus proves the importance of their role in complicated endocrine neurophysiological brain functions and in the regulation of the immune system. At the same time, the fact that vasopressin, oxytocin, PRPs, cardioactive protein-hormonal complex, and corticotropin-releasing factor are synthesized in the same hypothalamic nuclei as mentioned above[31], should be taken into consideration.

It is of primary importance to understand in which kind of cells (parvo- and magnocellular) and cellular compartments each of them is synthesized and what are the mechanisms of their interrelationship with classical neurotransmitters as well.

In the brain, microglia and astroglia have properties of immune cells. Microglia can be activated by various internal and external stimuli, resulting in cytokine expression. The inflammatory cytokines IL-1α and IL-1β, IL-4, IL-6, and TNF-α affect the secretion processes of the hypothalamus and neurohypophysis and thus activate the HPA axis[60,61].

The activity of immunocompetent cells can be modulated by oligo- and polypeptides and, conversely, ILs produced by activated immunocompetent cells affect the neurons and glial cells[55,62,63]. There is strong evidence for the existence of cytokine receptors in human neurons, astrocytes, and microglia and for a communication between the immune, endocrine, and central nervous systems mediated by peptide hormones and receptors common to these systems[52,64].

Cytokines such as IL-1, IL-2, IL-6, and also TNF-α, are able to modulate several functions of CNS neurons. IL-1 is of particular interest for elucidation of neurohormonal mechanisms of immune system regulation by the brain.

IL-1 is synthesized in the hypothalamus, and its level increases several times under stress[51,40]; it inhibits the immune system by activation of the HPA axis[68].

IL-2 regulates the effects of central dopaminergic neurons modulating their cholinergic, noradrenergic, serotoninergic, and glutaminergic functions. At the same time, IL-2 stimulates the secretion of vasopressin by rat hypothalamic cells in vitro[69].

IL-1β-like immunoreactive (ir) varicose fibers display a characteristic pattern within the hypothalamus. The densest accumulations of the fibers were found in periventricular regions that participate in the anterior pituitary control. Immunoreactive fibers were found throughout the periventricular and arcuate nuclei of the hypothalamus, as well as in the parvocellular part of the paraventricular nucleus. IL-1βir fibers could be traced into the infundibulum, including the region of the median eminence containing the hypophyseal portal vessels. In addition, IL-1βir innervation was seen in the magnocellular part of the paraventricular nucleus, and to a lesser extent in the supraoptic nucleus, among the cell bodies of neurons which secrete oxytocin and arginine vasopressin (AVP) from the posterior pituitary gland.

In view of the presence of IL-1-binding sites in the hypothalamus, the investigators speculate that IL-1 may induce CRF secretion by direct action in the hypothalamus, possibly on CRF neurons.

Our data indicate that IL-1 contained in the neurosecretory granules can produce an effect on the CRF neurons or on the hypothalamus-hypophyseal axis.

I suppose that these questions deserve special attention. Thorough study of the interrelationship of interleukins with the HPA gland axis will allow us to understand the role of these compounds in brain functions.

We made our first attempt to study the links and chains of relationship of interleukins, newly discovered cytokines (See Chapter II), and hormones (See Chapter III).

2

DISCOVERY OF NEW NEUROSECRETORY CYTOKINES: PROLINE-RICH PEPTIDES PRODUCED BY N. PARAVENTRICULARIS AND N. SUPRAOPTICUS OF HYPOTHALAMUS

2.1. ISOLATION OF PROLINE-RICH PEPTIDES FROM THE NEUROSECRETORY GRANULES OF BOVINE NEUROHYPOPHYSIS: PRIMARY STRUCTURES, MASS-SPECTRAL CHARACTERISTICS, AND LOCALIZATION.

Materials. The protease inhibitors phenylmethylsulfonyl fluoride, benzamidine, and leupeptin were obtained from Sigma. Acetonitrile (high-performance liquid chromatography (HPLC) grade) was obtained from Merck (Germany). Trifluoroacetic acid (TFA) (sequence grade) was from Knauer (Germany). Water was subjected to deionization using the Milli Q System (Millipore). All other chemicals used were of analytical grade.

Isolation and Purification of Secretory Granules. Neurosecretory granules were isolated from bovine neurohypophysis (Fig. 8) by a modified method described elsewhere[45,70,71]. The neurohypophyses were dissected out directly after the animals were slaughtered and then homogenized in 0.25 M sucrose (the tissue/buffer weight/volume ratio 1:10) in the presence of the protease inhibitors (5 mM phenylmethylsulfonyl fluoride (PMSF), 1 mM benzamidine, and 10 μM leupeptin). The procedures of granule preparation and purification were carried out in the medium containing the above-mentioned protease inhibitors at 4°C. Cell debris, membranes, and intact nuclei were removed by homogenate centrifugation at 1,000 x g for 10 min. Supernatant was centrifuged at 18,000 x g for 30 min to precipitate the organelles. The secretory granules in this fraction were separated from mitochondria, lysosomes, and microsomes by centrifugation through 1.6 M sucrose (100,000 x g for 90 min). After centrifugation the granule fraction was concentrated between the 1.6 and 2.0 M sucrose layers. The granules were pooled, suspended in the medium used for homogenization, and precipitated by centrifugation at 20,000 x g for 30 min.

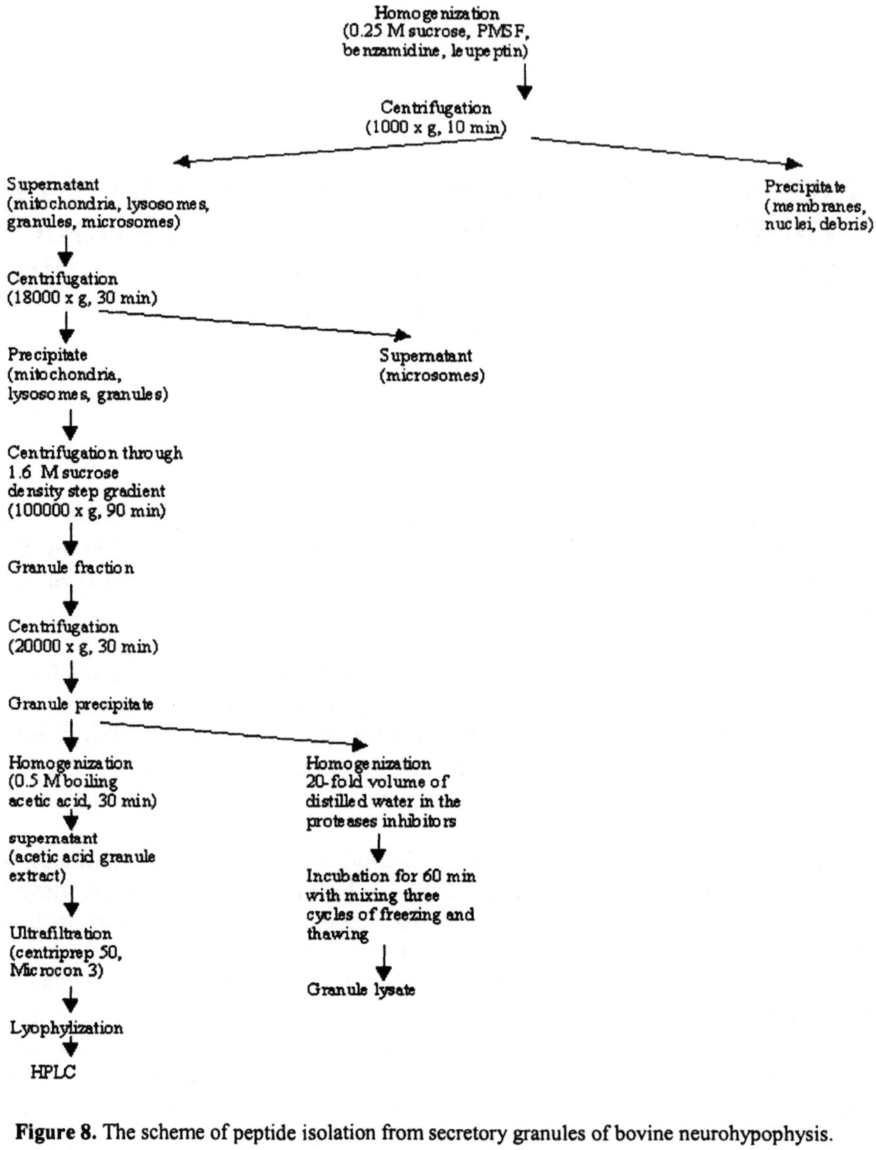

Figure 8. The scheme of peptide isolation from secretory granules of bovine neurohypophysis.

Preparation of the Granule Extract. Disruption of granules was performed by homogenization of granule precipitate in boiling 0.5 M acetic acid for 30 min. After centrifugation the extract was neutralized to pH 6.0, fractionated by means of the ultramembranes (Centriprep 50 and Microcon 3, "Amicon"), and lyophilized. The powder was then dissolved in 0.1% TFA, filtered through a 0.45 μ Sartorius filter, and the filtrate was subjected to reversed-phase HPLC.

Chromatographic Purification. The procedure was carried out using a Gilson HPLC system (France) on an Aquapore RP-300, Applied Biosystems (USA) C18 column (4.6 x

220 mm), equipped with a guard column (3.2 x 15 mm). Elution was accomplished using aqueous 0.1% TFA (buffer A) and 0.08% TFA in acetonitrile (buffer B), with a linear gradient from 0 to 33% buffer B for 60 min at a flow rate of 0.5 ml/min. In some cases rechromatography of the peaks was carried out using the Vydac (USA) C18 column (4.6 x 250 mm). Peaks detected by UV absorbance at 214 nm were collected and dried in a Speedvac drier (Savant, USA) (Fig. 9).

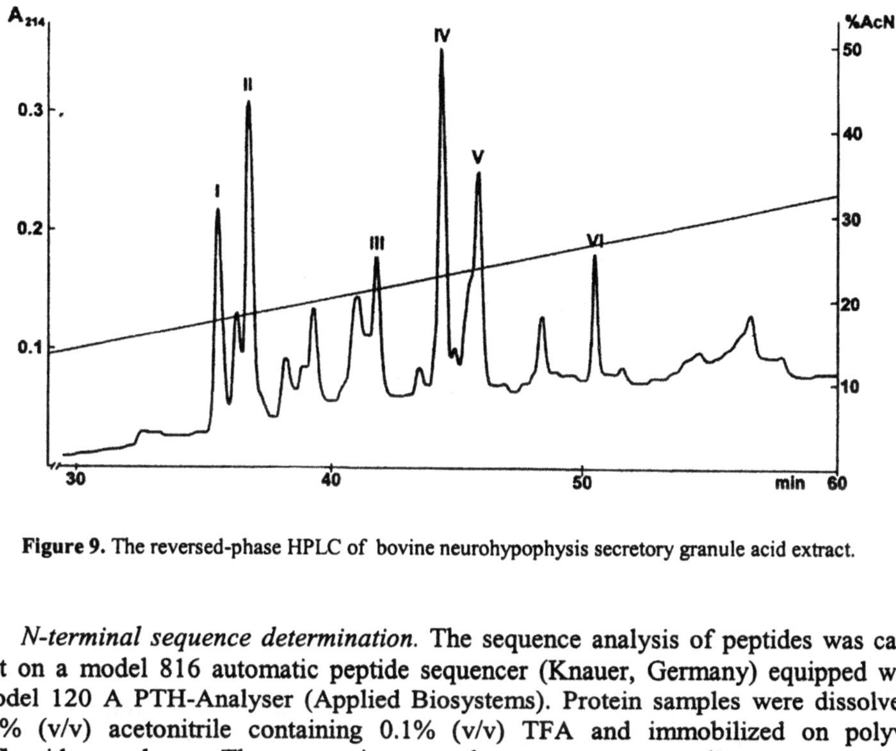

Figure 9. The reversed-phase HPLC of bovine neurohypophysis secretory granule acid extract.

N-terminal sequence determination. The sequence analysis of peptides was carried out on a model 816 automatic peptide sequencer (Knauer, Germany) equipped with a model 120 A PTH-Analyser (Applied Biosystems). Protein samples were dissolved in 50% (v/v) acetonitrile containing 0.1% (v/v) TFA and immobilized on polyvinyl difluoride membrane. The sequencing procedures were run according to a conventional program.

Using the RP300 C18 column we separated six major peptides (Fig. 9) from the neurohypophysis secretory granule acid extract fractionated previously by ultramembranes Centriprep 50 and Microcon 3. These peptides were eluted in the region from 17 to 27% buffer B, at approximately 17.3 (1), 18.1 (II), 21.5 (III), 23.1 (IV), 24.1 (V) and 26.9% (VI) of buffer B (Fig. 9). All of them were subjected to Edman microsequence analysis. Samples III and V were subjected to rechromatography on the Vydac C18 column before sequencing.

Analysis of amino acid sequences of the peptides (Table 2) showed that the peptides corresponding to peaks V and VI were vasopressin and oxytocin, respectively.

Table 2. The primary structure of neuropeptides isolated from neurosecretory granules of bovine neurohypophysis

PEAK	PEPTIDES		
I.	Ala-Pro-Glu-Pro-Ala-Glu-Pro-Ala-Gln-Pro	(27-36)	PRP-4
II.	Ala-Gly-Ala-Pro-Glu-Pro-Ala-Glu-Pro-Ala-Gln-Pro-Gly	(25-37)	PRP-3
III.	Ala-Gly-Ala-Pro-Glu-Pro-Ala-Glu-Pro-Ala-Gln-Pro-Gly-Val	(25-38)	PRP-2
IV.	Ala-Gly-Ala-Pro-Glu-Pro-Ala-Glu-Pro-Ala-Gln-Pro-Gly-Val-Tyr	(25-39)	PRP-1
V.	-H-Cys-Tyr-Phe-Gln-Asn-Cys-Pro-Lys-Gly-NH_2		(Vasopressin)
VI.	H-Cys-Tyr-Ile-Gln-Asn-Cys-Pro-Leu-Gly-NH_2		(Oxytocin)

The peptides corresponding to peaks I, II, III, and IV (Fig. 9) represented C-terminal fragments 27-36, 25-37, 25-38, and 25-39 of the neurophysin-vasopressin-associated glycoprotein (NVAG) [72,80]. These peptides were designated as PRPs (proline rich peptides) : PRP-4, PRP-3, PRP-2, and PRP-1, respectively.

In Fig. 10 the mass-spectrum of PRP-1 from bovine hypothalamus is shown. The peptide consists of 15 amino acid residues and has an apparent molecular mass of 1475.26 Da. In Fig. 11 the mass-spectrum of second polypeptide (PRP-2) is shown. The molecular mass of PRP-2 is 1311.

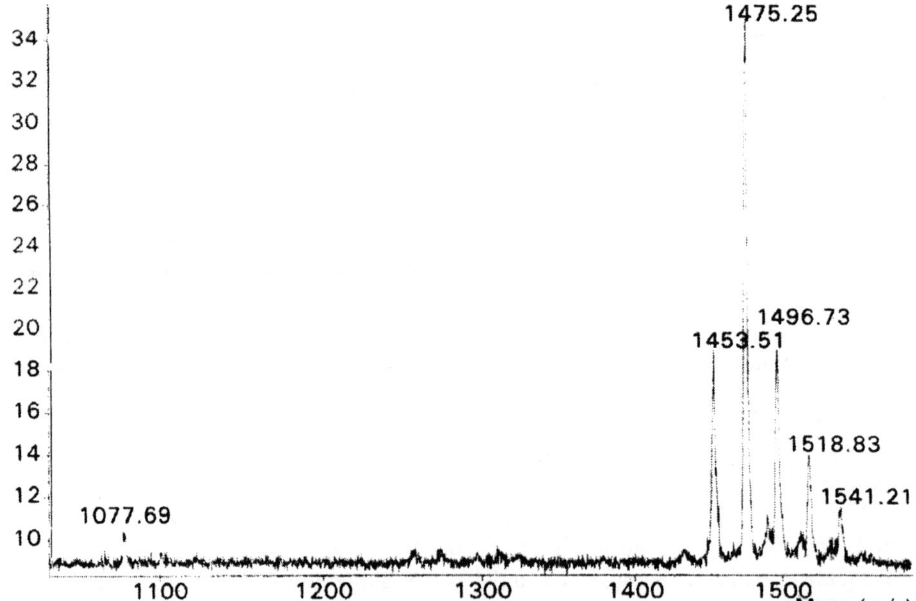

Figure 10. Mass-spectrum of PRP-1 (AGAPEPAEPAQPGVY). Mass Spectrometry (MALDI-TOF) was performed usingVISION 2000 mass spectrometer (Thermo Bioanalysis) with 2,5-dihydroxybenzoic acid as a matrix.

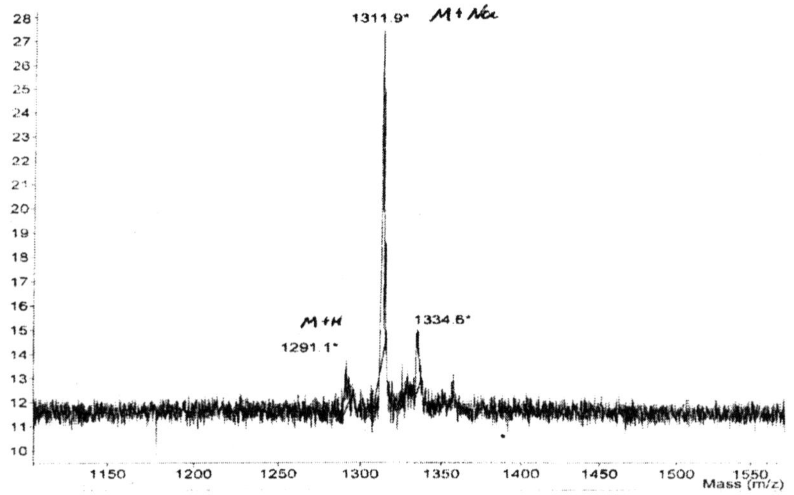

Figure 11. Mass-spectrum of bovine PRP-2 (AGAPEPAEPAQPGV). Mass Spectrometry (MALDI-TOF) was performed usingVISION 2000 mass spectrometer (Thermo Bioanalysis) with of 2,5-dihydroxybenzoic acid as a matrix.

As has been shown, the neurosecretory granules of the neurohypophysis also contain PRP, consisting of 13 amino acid residues with C-terminal glycine. In Fig. 12 is shown the mass-spectrum of the above mentioned polypeptide (PRP-3) with an apparent molecular mass 1193.3 Da.

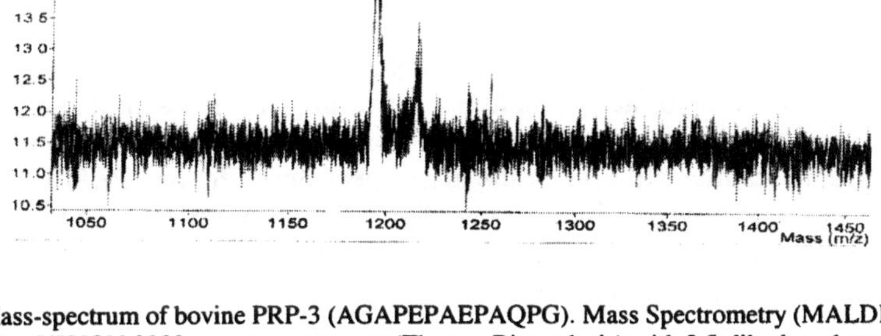

Figure 12. Mass-spectrum of bovine PRP-3 (AGAPEPAEPAQPG). Mass Spectrometry (MALDI-TOF) was performed using VISION 2000 mass spectrometer (Thermo Bioanalysis) with 2,5-dihydroxybenzoic acid as a matrix

It is of great interest that a peptide containing 10 amino acid residues with C-terminal proline was extracted from the neurosecretory granules of bovine neurohypophisis. The mass-spectrum of PRP-4 with a molecular mass of 1006.5 Da is shown in Fig. 13.

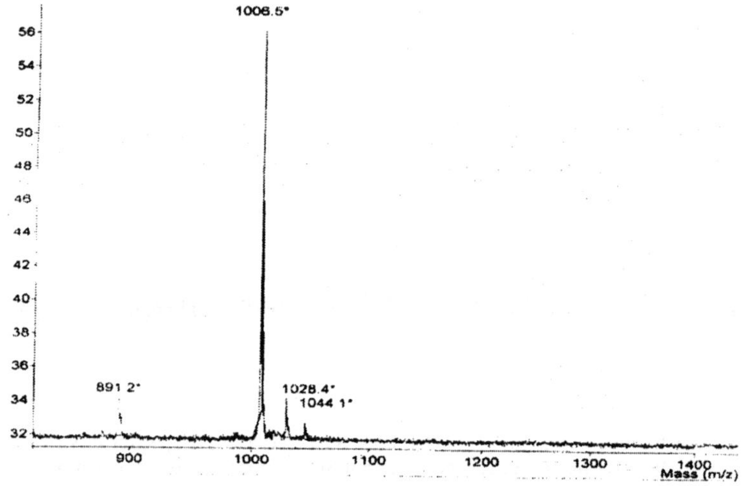

Figure 13. Mass-spectrum of bovine PRP-4 (APEPAEPAQP). Mass Spectrometry (MALDI-TOF) was performed using VISION 2000 mass spectrometer (Thermo Bioanalysis) with 2,5-dihydroxybenzoic acid as a matrix.

In my opinion this polypeptide has great importance for supporting the stability of cell membranes, in particular, membranes of erythrocytes. It appears to be highly active neuroprotector with immunomodulatory and antibacterial activity. The amidated form of the peptide can have a larger half-life than PRP-4 without C-terminal proline-amide.

Further studies of physico-chemical and biological properties of PRPs are in progress. At present we are work on chemical synthesis of amidated PRPs.

It is suggested that peptide PRP-4 is formed by protease cleavage of higher molecular mass PRPs. This occurs either in the body of neurosecretory cells of hypothalamus or during the axonal transport from hypothalamus to neurohypophysis. For formation of this 10-member polypeptide, a protease has to split the peptide between Gly-Ala (26-27 bond) and Pro-Gly (36-37 bond).

It was very interesting to study the physiological properties of human PRP which also consists of 15 amino acid residues. According to mass-spectral analysis the molecular mass of human PRP-5 is 1560.5 Da. This polypeptide differs from bovine PRP in three amino acid residues. In Fig. 14 the mass-spectrum of human PRP-5 is shown.

In the structure of bovine 15-amino acid PRP Ala is in a position against Phe-31 in the human peptide. In place of Gly-Val in bovine PRP, Asp-Ala is in human PRP (37-38). As it is shown below, the positions of proline residues in human and bovine PRPs are the same :

Bovine PRP-1 : Ala-Gly-Ala-Pro-Glu-Pro-Ala-Glu-Pro-Ala-Gln-Pro-Gly-Val-Tyr
Human PRP-5 : Ala-Gly-Ala-Pro-Glu-Pro-Phe-Glu-Pro-Ala-Gln-Pro-Asp-Ala-Tyr.

In 1985 D. Richter by the method of gene engineering determined the primary structure of hypothalamic NVAG in as a continuous molecule. NVAG is formed in NPV.

The nucleotide and amino acid sequences NVAG as a precursor form of rat and calf vasopressin and also the possible mechanisms of fragmentation have been discovered[72]. In Table 3 the complete amino acid sequence of bovine NVAG is shown.

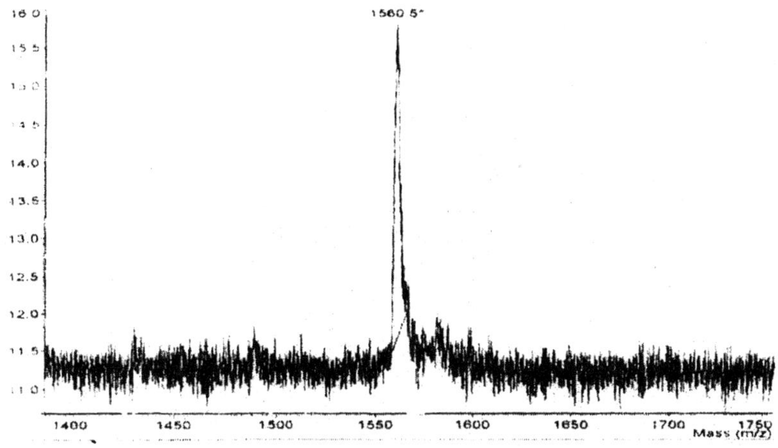

Figure 14. Mass-spectrum of human PRP-5 (AGPEPFEPAQDAY). Mass Spectrometry (MALDI-TOF) was performed using VISION 2000 mass spectrometer (Thermo Bioanalysis) with 2,5-dihydroxybenzoic acid as a matrix.

Table 3. Aminoacid sequence of the bovine vasopressin precursor (Richter, 1985).

Met-Pro-Asp-Ala-Thr-Leu-Pro-Ala-Cys-Phe-Leu-Ser-Leu-Leu-Ala-Phe-Thr-Ser-Ala

Vasopressin ... Neurophysin ... +20

Cys-Tyr-Phe-Gln-Asn-Cys-Pro-Arg-Gly Gly-Lys-Arg Ala-Met-Ser-Asp-Leu-Glu-Leu-Arg-Gln-Cys-Leu-Pro-Cys

+40

Gly-Pro-Gly-Gly-Lys-Gly-Arg-Cys-Phe-Gly-Pro-Ser-Ile-Cys-Cys-Gly-Asp-Glu-Leu-Gly-Cys-Phe-Val-Gly-Thr-Ala-Glu

+60

Ala-Leu-Arg-Cys-Gln-Glu-Glu-Asn-Tyr-Leu-Pro-Ser-Pro-Cys-Gln-Ser-Gly-Gln-Lys-Pro-Cys-Gly-Ser-Gly-Gly-Arg-Cys

+80 ... +100

Ala-Ala-Ala-Gly-Ile-Cys-Cys-Asn-Asp-Glu-Ser-Cys-Val-Thr-Glu-Pro-Glu-Cys-Arg-Glu-Gly-Val-Gly-Phe-Pro-Arg-Arg

Glycoprotein ... +120

Val Arg Ala-Asn-Asp-Arg-Ser-Asn-Ala-Thr-Leu-Leu-Asp-Gly-Pro-Ser-Gly-Ala-Leu-Leu-Leu-Arg-Leu-Val-Gln

+140

Leu-Ala-Gly-Ala-Pro-Glu-Pro-Ala-Glu-Pro-Ala-Gln-Pro-Gly-Val-Tyr

Thus, bovine preprovasopressin consists of 168 amino acid residues, and its molecular mass is 17310 Da. NVAG of rat also consists of 168 amino acid residues having molecular mass of 17826 Da (Table 3). The glycosylation site Asn-Ala-Thr and the leucine-rich central part are well conserved in all species so far known. The consecutive leucine residues may represent native processing signals for converting the glycoprotein into subfractions[80]. Although the glycoprotein in the composition of NVAG is localized in the vasopressin-producing magnocellular neurons, however the biological role of this glycoprotein remains to be detalized[73]. Cohen's group has accumulated considerable materials about the presence of high-molecular forms of neurophysin and somatostatin precursors with molecular mass of 17 kDa and 20 kDa, respectively, in mouse hypothalamus. The authors suggest that this protein also contains carbohydrates[74].

In 1981 P. Cohen and his group reported about the presence in bovine neurohypophysis of neurophysins with molecular mass of 80 kDa. According to their data, the degradation of 80 kDa protein results in formation of 68 Da and 10 kDa fragments[75]. But the authors did not study whether these proteins were related to neurophysin I and neurophysins II.

Analysis of the nucleotide sequence of cloned cDNA that encodes bovine arginin-vasopressin-neurophysin precursor confirmed that the glycopeptide region is at the C-terminus of the precursor[76-80]. The glycopeptide was isolated from bovine, sheep, rat, and human neurohypophyses[80,81,82] as well as from porcine hypothalamus, and its prolactin-releasing role was shown (Table 4).

Table 4. The complete sequence of the isolated human pituitary glycoprotein. A comparison with the sequences reported for the homologous pig, sheep and ox peptides is also shown [Smith, Massoy, 1979; Seidah et al., 1981].

```
                  CHO
                   |
        1                  10        15         20         25        30        35       39
Human   Ala-Ser-Asp-Arg-Ser-Asn-Ala-Thr-[Gln]-Leu-Asp-Gly-Pro-[Ala]-Gly-Ala-Leu-Leu-Leu-Arg-Leu-Val-Gln-Leu-Ala-Gly-Ala-Pro-Glu-Pro-Phe-Glu-Pro-Ala-Gln-Pro-[Asp-Ala]-Tyr

                  CHO
                   |                                                                                                                                         39
PIG     Ala-Ser-Asp-Arg-Ser-Asn-Ala-Thr-[Leu]-Leu-Asp-Gly-Pro-[Ser]-Gly-Ala-Leu-Leu-Leu-Arg-Leu-Val-Gln-Leu-Ala-Gly-Ala-Pro-Glu-Pro-[Ala]-Glu-Pro-Ala-Gln-Pro-[Gly-Val]-Tyr

                  CHO
                   |
SHEEP   Ala-Ser-Asp-Arg-Ser-Asn-Ala-Thr-[Leu]-Leu-Asp-Gly-Pro-[Ser]-Gly-Ala-Leu-Leu-Leu-Arg-Leu-Val-Gln-Leu-Ala-[Ala]-Ala-Pro-Glu-Pro-[Ala]-Glu-Pro-Ala-Gln-Pro-[Gly-Val]-Tyr

                  CHO
                   |
OX      Ala-[Asn]-Asp-Arg-Ser-Asn-Ala-Thr-[Leu]-Leu-Asp-Gly-Pro-[Ser]-Gly-Ala-Leu-Leu-Leu-Arg-Leu-Val-Gln-Leu-Ala-Gly-Ala-Pro-Glu-Pro-[Ala]-Glu-Pro-Ala-Gln-Pro-[Gly-Val]-Tyr
```

It is suggested that the four novel proline rich peptides isolated by us are formed naturally from the glycoprotein. Formation of the glycopeptide fragments during isolation is unlikely because all the procedures of the secretory granule isolation and purification were performed in the presence of the protease inhibitors, and the granules were extracted by boiling acetic acid.

Other peptides derived from C- or N-terminus of the glycopeptide were identified in porcine and rat hypophysis extracts[80,81,82]. Thus, four different naturally occurring fragments of the glycoprotein have also been isolated[80,82]. Existence of two fragments containing the amino acid residues 1 to 10 and 1 to 19, indicate that the hypophysis contains proteases, which act at the C-side of paired leucine residues. A similar fragment which corresponds to residues 1-18 of the glycopeptide was found in hypophysis powder[78], which indicates the same protease cleavage specificity. Formation of two other peptides, which account for residues 23-39 and 26-39 of the glycopeptide, involves the cleavage on the carboxyl side of Val-Gly and Ala-Gly bonds, respectively[80]

Four peptides derived from the C-terminus of the NVAG were also isolated from rat neurohypophysis[82]. They were identified as the fragments of the glycoprotein 1-19, 1-20, 22-37, and 22-39.

Thus, processing of the glycopeptide could produce peptides specific at post-Arg and PRP-1-post-Leu cleavages. Leu-residue serves as a processing signal for an endopeptidase with chymotrypsin-like specificity[83]. Leu-Xaa cleavages, with Xaa being a hydrophobic or neutral amino acid, play a role in the formation of relaxin[84], atrial natriuretic factor[85], γ-endorphin[86,87], angiotensin I[88], and other hormones. Besides, a specific role for subtilisin-like prohormone convertase (SPC3) in the processing of NVAG in vitro was shown[89,90]. This mechanism could also be responsible for the formation of the peptides isolated by us from the neurohypophysis secretory granules.

It was established that the glycoprotein is a releasing factor for prolactin[81]. The effect of the glycoprotein (39 amino acid residues) on hypophysis cell monoclonal culture was studied, and it was shown that the glycopotein at nanomolar concentrations stimulates the secretion of prolactin into the culture medium. Antiserum against the glycoprotein abolishes the effect.

The presence of glycoprotein in blood indicates the splitting of Arg-Ala bond. For the formation of 25-39 polypeptide, the bonds Leu-Ala in sites of 27-36, 25-37, and 25-38 must be split successively. The following peptide bonds have to be enzymatically split for the formation of individual peptides during the transport of neurosecretory granules from hypothalamus to hypophysis: Gly-Ala and Gly-Pro (27-36), Leu-Ala (25-39), Leu-Ala and Pro-Gly (25-37), Leu-Ala and Val-Tyr (25-38).

Back in 1987 we showed that catepsin B of brain splits at pH 5 Ala-Leu bond in the structure of chromofore oligopeptides[91]. There are Gly-Cly, Ala-Gly, and Ala-Leu aminopeptidases in the brain. According to our data, cathepsin B of brain (EC 3.4.22.1) acts as an endopeptidase at pH 6, it splits Val-Tyr bond in the structure of angiotensin II. Cathepsin-like proteases take part in the processing of proteins and precursors of peptide hormones[91,92]. Serine-endopeptidase/proline-endopeptidase (EC 3.4.21.26), cytosol-peptidehydrolase of brain, splits Pro-Gln bond in P substance (slow splitting). It is probable that the above mentioned enzymes are responsible for processing of the proline rich peptides. In any case, the presence of these polypeptides, without doubt, indicates the presence of proteolytic enzymes in neurosecretory cells, and probably in neurosecretory granules[93].

B. Pickering and coworkers[93] supposed that the synthesis of provasopressin and prooxytocin takes place in ribosomes, and they are incorporated into the granules together with proteolitic enzymes. During passage of the granules along the axons to their storage sites in the nerve terminals provasopressin is cleaved to give vasopressin and neurophysin.

In the neurosecretory granules of mammals protein phosphorylation is catalysed by the endogenous protein kinase[94]. ATP stimulates the release of vasopressin from isolated neurosecretory granules[95]. cAMP is involved in the hormone release in the hind part of the hypophysis[96]. The transport of neurosecretory granules is inhibited by colchicine [97].

Recently it was shown, that after removal of the neurohypophysis, prolactin secretion is blocked which indicates the presence of an unidentificated prolactin-releasing factor in the neurohypophysis [81].

Experiments demonstrated that the addition of the glycoprotein to hypophysis cell monolayer culture stimulates the secretion of prolactin to a similar degree as does thyreotropin-releasing hormone (TRH). The same concentrations of the glycoprotein do not produce any effect on the secretion of luteinizing-, adrenocorticotropin-, and thyroid-stimulating hormones, and also on growth hormone.

For clearing up the role of glycoprotein in secretion of prolactin *in vivo,* lactating rats were intravenously injected with an antiserum to the glycoprotein.

The level of prolactin was studied by radioimmunological method. After injection of antiserum in blood, the level of prolactin is significantly lower than in control rats, which were injected the same amount of normal rabbit serum. There was also deep and long inhibition of milk secretion under suckling in three females, which were injected with the antiserum. After one week, the quantity of milk was 192 ± 13.4 g, and it was 244.7 ± 12.3 g in the case of four control animals. Antiserum against glycoprotein did not crossreact with peptides of neurohypophysis: ovine glycopeptide, vasopressin, oxytocin, neurophysin I and II, vasoactive intestinal peptide, angiotensin III, dinorphin A, atrial natriuretic factor, and thyrotropin releasing factor.

The disturbance of NPV of hypothalamus or a surgical removal of neurohypophysis blocked the release of prolactin induced by suckling in rats.

The prolactins of different mammalia have similar biological and immunological properties, and their synthesis occurs by a known pathway – through a prohormone. The transcription of prolactin gene is regulated by factors, which also regulate the hormone secretion.

The stimulators of prolactin secretion are pregnancy, estrogen, sleep, stress, TRH, dopaminergic antagonists of opioids, serotonin, antagonists of histamine, adrenergic antagonists. The inhibitors of prolactin secretion are dopamine, dopaminergic antagonists, prolactin, and GABA. During pregnancy the content of prolactin increases in blood serum 20-fold.

Apparently, that is the result of an increase in estrogen content, which provokes hyperplasia of prolactin-producing cells and synthesis of the hormone, inducing the gene transcription.

It has been established, that vasoactive intestinal polypeptide (VIP), like TRH, also appears to be a prolactin-releasing factor. As is known, VIP is present in high concentrations in hypothalamus and portal vessels and serves as a powerful stimulator of prolactin secretion. The physiological significance of positive effect of opioid peptides and serotonin, and also negative effect of noradrenaline, on prolactin secretion, remain indecisive.

It is of interest if PRP-1 participates in the biosynthesis of prolactin and its release from the hypophysis into blood.

As will be demonstrated below (Chapter III, Table 5), the content of prolactin in the brain is 1.96±3.8 ng/ml. Under aluminium neurotoxicosis the content of prolactin in brain homogenate increases to some extent (2.9±0.9 ng/ml). Meanwhile, under simultaneous administration of PRP-1 in doses of 15 μg and 1.5 μg the content of prolactin increases 15 and 8 times, respectively. Such a regularity is observed in changes of the prolactin content in blood serum of rats. Thus, it can be supposed that PRP-1 participates in the regulation of prolactin synthesis and release.

Sometimes lactation without pregnancy is observed. In men, the results of a surplus of prolactin are the decrease in testosterone secretion and sperm formation. The content of vasopressin in the blood increases 50 times causing hemorrhage, pain, or nausea.

Hypophysis removal blocks the secretion of prolactin during suckling[98]. Interestingly, the effect of the glycoprotein on prolactin secretion is not abolished in the presence of prolactin-inhibiting dopamine.

It is interesting to note that neurophysin, which is oxytocin precursor, does not contain the above mentioned glycoprotein in its C-terminal region.

The neurophysins appear to be typical cellular polyproteins. Until the present time our knowledge about functions of both the glycoprotein and neurophysin is insufficient.

It is obvious, that NVAG within neurosecretory granules slowly moves to neurohypophysis by axonal transport via hypothalamo-neurohypophyseal nervous tract, where the separation of oxytocin and vasopressin from NVAG as well proline rich peptides takes place.

However, as we show it, neurophysins and their structural components play exclusively important role in the functions of the organism, both in endocrine and immune functions of the brain and peripheral immune organs. For understanding of the functions of neurohormones it is enough to follow the ways of their transport in the organism.

From NPV where the basic site of neurophysin synthesis is, the fibers extend as far as rostracaudal segment of spinal cord.

About 1500 neurons from each side of NPV of hypothalamus project as for as NTS, dorsal motor nucleus of vagus (DMN), and spinal cord. Monosynaptic effector way leads from NPV as for as DMN [99].

Hormones, which are formed by neurosecretory cells of hypothalamus, play a role in regulation of functions of adenohypophysis, brain, and CNS.

The data exist about the presence of direct projections from central nucleus of amigdala as far as NPV of hypothalamus, and also in lateral septum and midbrain periaqueductal gray matter[100]. Neurophysin II was detected also in human epiphysis[101].

These data show that the other possible way of neurophysin formation is the transport from hypothalamic nuclei[102].

2.2. IDENTIFICATION OF PRP-1 IN THE BRAIN BY IMMUNOHISTOCHEMICAL METHODS USING POLYCLONAL ANTIBODIES

In the intact rat tisues PRP-1-immunoreactivity (PRP-1-IR) was found in the magnocellular neurons of the NSO and in parvocellular neurons concentrated in the rostral and medial subdivision of the NPV (Fig. 15 A, B). Daily administration of PRP-1 into intact animals for 3 weeks induces significant activation of PRP-1 in the brain microstructure (Fig. 15 C-F). PRP-1 immunoreactivity was much stronger in the NSO in comparison with that in the NPV. Besides, in the periventricular hypothalamic region some single neurons showed PRP-1-IR. Other brain regions containing PRP-1-

immunoreactive cells were certain limbic areas such as the amygdaloid nuclei and hippocampal complex, as well as cell groups in the medulla oblongata: hypoglossal and facial nuclei, nucleus of the solitary tract, olives, and reticular formation[103].

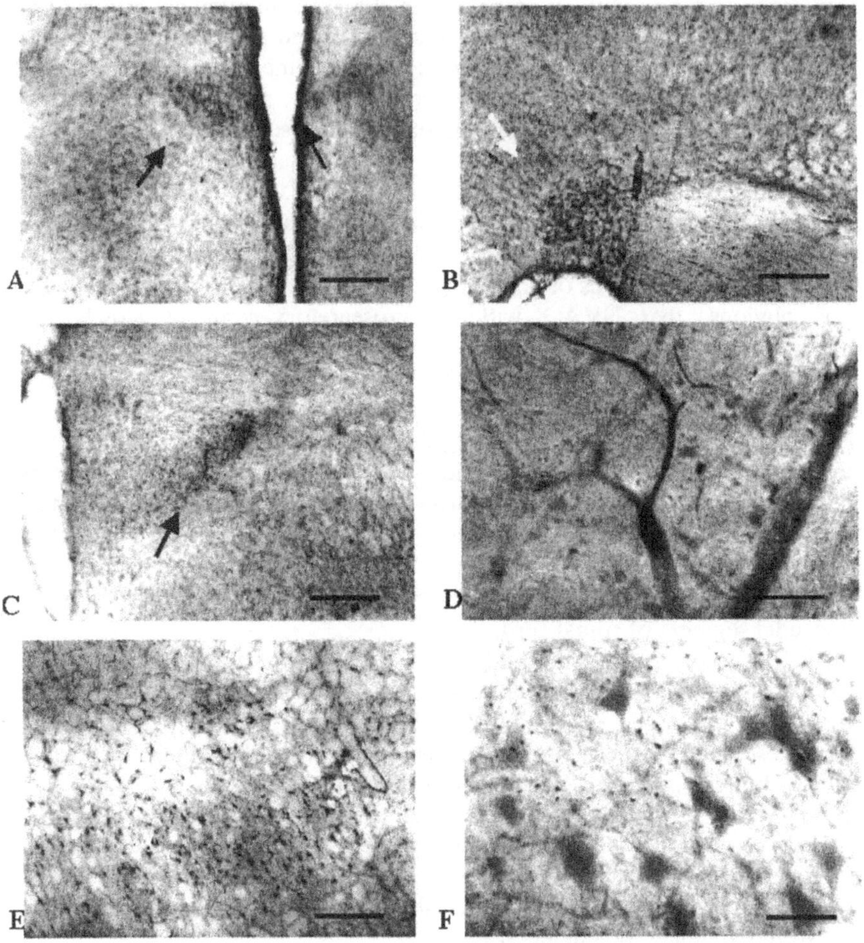

Figure 15. PRP-1-IR in the intact rat hypothalamus (A, B): PRP-IR of parvicellular neurons in (A) NPV (arrows) and magnocellular ones in (B) NSO (white arrow). PRP-1-IR in the brain of the PRP-1 treated rats (C-F): increased number of PRP-1-IR neurons in the NPV (C); PRP-1-IR (D) of blood vessel and (E, F) neurons in the facial nucleus. Bars: 50 μm (A, B, C, E); 25 μm (D, F).

In rats exposed to trauma stress 3 weeks after spinal cord (SC) hemisection, an asymmetry in the PRP-1 localization was observed in the NPV and NSO. In addition, some PRP- immunoreactive varicose nerve fibers were seen in the NPV and NSO.

In rats with SC hemisection, PRP-1 administration induces significant activation of PRP-1-IR and increases the number of PRP-1-immunoreactive nerve cells and fibers in the NPV (Fig. 16 C,D) and NSO (Fig. 16 E). Strong PRP-1 immunoreactivity of neurons was observed in the periventricular hypothalamic area, amygdaloid nuclei (Fig. 16 E), and median eminence (Fig. 16 F). Apart from these changes, the asymmetry in PRP-1 localization observed in control rats (without PRP-1 treatment), disappeared after PRP-1 injection (Fig. 16 C). Data on significant redistribution of PRP-1 in other brain regions are not presented here.

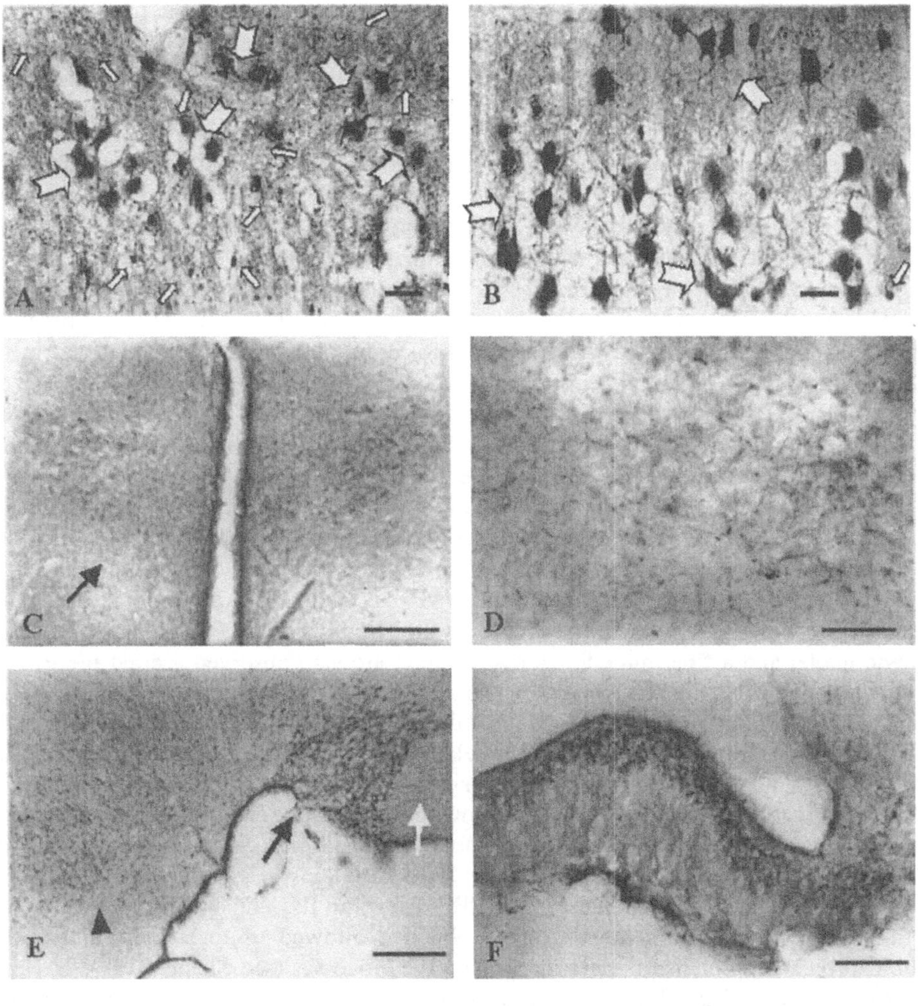

Figure 16. PRP-1-immunoreactivity in the neural structures of rats with the SC hemisection following the PRP-1 treatment. (A) In the SC PRP-1-IR gliocyte nuclei (small arrows) and fine reaction product are seen around the immunopositive degenerative MN (large arrows) situated in the lightened pericellular area. (B) Heavy PRP-1-IR regenerated MN and processes in the injured rats after the PRP-1 administration. (C, D) PRP-1-IR neurons and increased number of nerve fibers in NPV. Strong PRP-1-IR neurons in (E) amygdaloid nucleus (arrow head) and in NSO (black arrow) as well as in (F) median eminence. Bars: 50 μm (C, E); 25 μm (A, B, D, F).

The appearance of PRP-1-immunoreactive nerve fibers in the NPV and the intensification of PRP-1-IR in certain cell groups of medulla oblongata and also in amygdaloid nucleus gives the basis to suggest that these projections could be raised from the above mentioned brain regions. This suggestion is based on literature data, which give evidence that amygdala and a majority of medullary cell groups project to the NPV[104,105]. It is also known that the median eminence has a key role in the stress response, constituting a link between the central nervous system and the pituitary gland. The median eminence receives axons from various sources most of which are intrahypothalamic (mainly from the NPV, NSO, and periventricular nucleus)[106,107].

In the median eminence, a high number of nerve cells and fibers show strong immunostaining for PRP-1 in rats with SC hemisection following PRP-1 injection (Fig. 16 F). Several neuropeptides are present in nerve terminals in the NPV that may function as neurotransmitters[108]. It is likely that nerve terminals containing PRP-1 originate in the above brain regions, reach the NPV, influencing PRP-1 synthesis and/or its axonal transport from the nucleus to the median eminence. Various neuropeptides, such as vasopressin, oxytocin, angiotensin II, VIP, corticotropin-releasing factor (CRF), and biogenic amines are present in the hypothalamus, mainly in the NPV, where the concentration of these neuropeptides is found to be 50-100 times lower than that in the median eminence.

Thus, PRP-1 was considered to be involved in the mechanism of response to stress (in this case, trauma-induced stress), and to act as a neurotransmitter and neuromodulator.

We have demonstrated the effect of the PRP-1 on the electrical activity of rat SC subjected to SC hemisection. Recently, an immunohistochemical study was carried out by us concerning the effect of PRP-1 on trauma-damaged SC, that resulted in recovery phases and growth of nerve fibers in SC, as well as glia proliferation and preservation of motoneuron (MN) integrity. Degenerative PRP-1-positive MN are found in animals subjected to SC hemisection without PRP-1 administration, where PRP-1-mmunoreactive gliocyte nuclei and a fine immunoreactive product are also observed around the PRP-1-immunopositive MN (Fig. 16 A). It is worth mentioning that the pericellular area is light (brain edema), which, perhaps, is a result of disturbed water-salt metabolism. After daily PRP-1 administration to animals for 3 weeks regenerated PRP-1-immunoreactive MN with processes are observed (Fig. 16 B). Besides, the light pericellular area decreases. In addition, an increased number of PRP-1-immunoreactive interneurons (IN) was found in rat SC under abovementioned experimental conditions. The presence of the monosynaptic pathways from the NPV to the IN is well known. It is possible that the MN recovery is associated with the increase of PRP-1 level in IN. The latter are considered to be the stabile structures in the SC. Data obtained allowed us to suggest that PRP-1 exhibits protective properties in neurodegenerative processes (see Chapter 6).

We also showed that PRP-1 injection resulted in the increase in NPV- and Iph-IR in intact rat brain neuronal structures. In contrast, PRP-1 inhibited the biosynthesis and release of Iph in the injured rats. Data obtained indicate the existence of a very close interrelationship between the PRP-1 and the above listed peptides. The functional significance of these interactions will be studied.

2.3. DETERMINATION OF PRP-1 IN THE THYMUS, LYMPH NODES, AND HEART USING MONOCLONAL ANTIBODIES

Taking into account the neuronal origin of PRP-1, it was important to study its distribution in the thymus, lymphatic nodes, bone-marrow hematopoietic cells, etc.

We will briefly describe below immunohistochemical methods for studying the localization of PRP-1 using monoclonal antibodies[109].

Monoclonal antibodies were obtained in Prof. V. Bedian's laboratory (Department of Molecular Biology and Genetics of the University of Pennsylvania).

Preparation of monoclonal anti-PRP-1 antibodies. Immunization. Antigens were prepared in sterile calcium-magnesium-free phosphate-buffered saline (PBS, 8.0 g NaCl, 0.2 g KCl, 1.15 g Na_2HPO_4, 0.2 g KH_2PO_4 per liter). Six-week-old BALB/c mice (Jackson Laboratories, Bar Harbor, ME) were used for immunizations. The primary immunization was done intraperitoneally (i/p) with the antigen mixed 1:1 with Titermax adjuvant (Sigma Chemical Co.) according to manufacturer's instructions. Subsequent immunizations were performed via subcutaneous (s/c) and i/p routes in alternating order, using Freund's incomplete adjuvant. Immunization dose was 30-100 μg per animal. Three days prior to taking the spleen for fusion a final boost of 50 μg was administered intravenously (i/v) without adjuvant. Iph and PRP-1 were mixed in equal weight and administered to a second mouse as a combined antigen, but hybridomas were screened against each component. Tail bleeds from immunized mice were used to determine serum titers. Detailed protocols are described by Bedian et al.[110].

Myelomas and lymphocytes. Sp2/0 Ag14 mouse myeloma line (ATCC) was used as the fusion partner. Sp2 cells were cultured in HY medium: 90% high glucose DMEM, 10% NCTC135 (both from Sigma) supplemented with 10% fetal bovine serum (FBS), and 2 mM L-glutamine (Gibco, Grand Island, NY). The cells were maintained in a density range of 0.1-1 million cell/ml. Fusion medium was composed of HY supplemented with 20% FBS, 5% hybridoma cloning factor (Fisher), 4 mM L-glutamine, 0.15 mg/ml oxaloacetate, 0.05 mg/ml pyruvate, 0.2 U/ml insulin, 13.6 mg/ml hypoxanthine, 1.0 mg/ml azaserine, 100 U/ml penicillin, 100 mg/ml streptomycin (Sigma). After administration of appropriate anesthetic, the mouse was killed, and the spleen was dissected and homogenized using a sterile "Cellector" stainless steel sieve homogenizer (Bellco, Vineland, NJ) with a 60-mesh screen. The cells were centrifuged at 300 g for 15 min. The pellet was resuspended in 5 ml of ice-cold erythrocyte lysis buffer (0.17 M NH_4Cl, pH 7.5), incubated on ice for 8 min, then 10 ml of HY was added, and the centrifugation was repeated. For in vitro stimulation with antigen, the cells were suspended in hybridoma culture/cloning medium at a density of 107 cell/ml, sterile antigen and adjuvant peptide were added to final concentrations of 1 μg/ml and 20 μg/ml, respectively, and incubated in a humidified 8% CO_2 incubator for 5 days prior to fusion.

Fusions. Fusion procedures were adapted from Lane et al. For every 10 million freshly harvested or in vitro stimulated lymphocytes, 6 million Sp2 cells from log phase cultures were added. Cells were washed in serum-free HY and pelleted in a 50 ml tube. One ml of PEG solution (45% PEG, 1.5-3 kDa range, 5% DMSO, 50% HY) prewarmed to 37°C was added over 30 s with mixing and gradually diluted with 12 ml of HY over the next 90 s. Cells were allowed to rest for 10 min, pelleted, resuspended in fusion medium, and distributed, 0.13 ml per well, into 96-well plates using a multichannel pipettor. Plates were incubated at 37°C under 8% CO_2, fed with 0.1 ml of hybridoma culture/cloning medium on day 7. Yellow supernatant was harvested from hybridoma-containing wells around day 10.

ELISA screening. Antigen-binding buffer was composed of 2.93 g $NaHCO_3$ and 1.59 g Na_2CO_3 per liter. Other buffers, such as 50 mM sodium borate buffer, pH 8.5, or PBS can also be used if binding in bicarbonate buffer is not efficient. ELISA plates were Nunc Maxisorb or Becton-Dickinson Probind (Fisher). It is worth trying both plates during assay development and selecting the one that shows stronger binding of specific antigen. Blocking solution can be 0.2-2.0% BSA or ovalbumin in PBS, or 0.1-2.0% instant milk. Wash buffer was PBST. Affinity-purified HRP and FITC-conjugated goat anti-mouse IgG (secondary antibodies) were from BMB (Indianapolis, IN). Substrate was freshly prepared from ABTS tablets (BMB). Fresh formaldehyde was prepared by dissolving 0.4 g of paraformaldehyde in 1 ml of 1 M NaOH at 60°C, neutralized with 1 M HCl, then 0.5 ml of 10X PBS was added, and the volume was adjusted to 10 ml. This solution should be kept tightly capped at 4°C and used within a few days.

Screening. Hybridoma supernatants were tested by ELISA on Nunc Maxisord plates. Antigen was dispensed (0.05 ml at 5 μg/ml in 50 mM sodium carbonate-bicarbonate buffer, pH 9.5) into wells and incubated overnight at 4°C. Wells were blocked with 1% BSA in PBS for 30 min at room temperature (RT). Supernatants were used without dilution and incubated 45 min at RT. Plates were washed 3 times, 10 min each, with PBST (PBS with 0.1% Tween 20). Plates were incubated 30 min at RT with 1 : 1000 dilution of HRP-conjugated anti-mouse antibody (BMB). After repeating the wash steps, 0.2 ml of fresh solution of substrate (ABTS, BMB) was added, and the absorbance was monitored using a multiplate reader with a 405 nm filter. Positive hybridomas were subcloned by limited dilutions and retested. Positive wells were visually scored as containing single clones which were expanded and frozen.

Immunohistochemical analysis of PRP-1 distribution. The investigations were carried out on heart atrial tissues removed from corpses of practically healthy persons who died from traumas and were subjected to autopsy within first hours after death. We used also rat thymus and lymph nodes.

Immunohistochemical analysis by peroxidase-antiperoxidase (PAP) reaction was performed on 5 μm cryostat sections. The sections were air-dried for 60 min and fixed in fresh acetone for 10 min at room temperature and were washed in PBS after fixation and after each incubation. Endogenous peroxidase activity was blocked by incubation of the sections for 30 min in PBS/3% hydrogen peroxide, and the slides were washed twice in PBS. The sections were preincubated for 30 min with 10% normal horse serum to avoid nonspecific immunoreactivity. After decanting the preincubation fluid the sections were incubated for 50 min at room temperature with the monoclonal anti-PRP-1 antibody as a primary antibody, at a 1 : 50 dilution. Afterwards they were washed 3 times for 5 minutes with PBS. Horseradish peroxidase-labeled rabbit anti-mouse immunoglobulin was used as a second-step reagent. The peroxidase label was detected with 3,3-diaminobenzidine tetrahydrochloride (Loba Chemie Wien-Fischamend) as a chromogen, together with hydrogen peroxide, for 30 min at room temperature. The sections were counterstained with hematoxylin. The control sections were not treated with monoclonal antibodies and conjugate.

Human heart. Right atrium. The study of PRP-1-IR in the tissues of human right atrium indicates that all layers of wall were negative. No staining was detected with anti-PRP-1 antibodies in the endocardial, myocardial, and pericardial cells. The reaction of the cell types that show diffuse and granular immunoreactivity for PRP-1 was observed in the cytoplasm of some neurons in the region of synoatrial node, within the network of muscle fibers (Fig. 17).

Rat thymus. PRP-1-IR was found in the cells of the thymic capillaries (Fig. 18). Deep in the medulla, where the epitheliocytes form a bulky cord, some of Hassell's corpuscles were positive (Fig. 19).

Rat lymph node. There was no PRP-1-positive reaction of the cells of capsule, follicles, as well as medullary cords. Specific reaction to antibodies was found with dark brown-stained mast cells, accumulated on the surface of cortical sinuses (Fig. 20a).

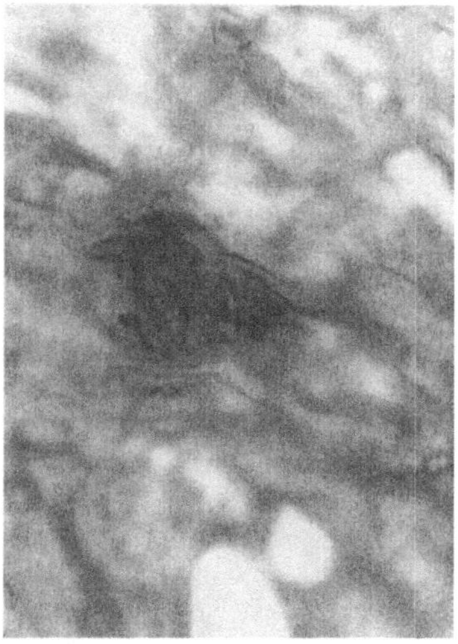

Figure 17. Micrograph from the area of sinoatrial node. Neurons with dark brown strongly positive reaction are seen between the whorled network of muscle fibers. PRP-1 immunostaining of human right atrium myocardium with anti-PRP-1 monoclonal antibody. PAP rection. Magnification x 1000.

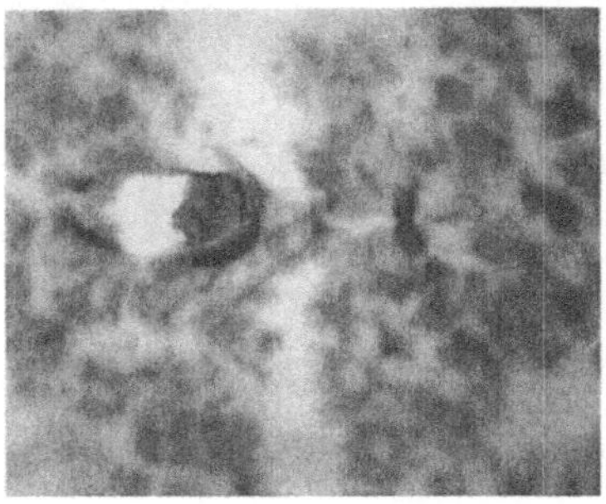

Figure 18. Positive reaction for PRP-1 in thymus vessels. PAP reaction. Magnification x 400.

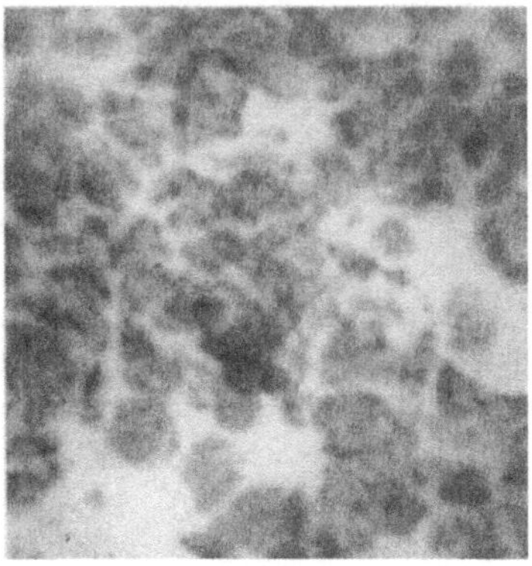

Figure 19. PRP-1 immunostaining of intact rat thymus. The medullar cells are negative, some of the Hassell's cells are positive. PAP reaction. Magnification x 400.

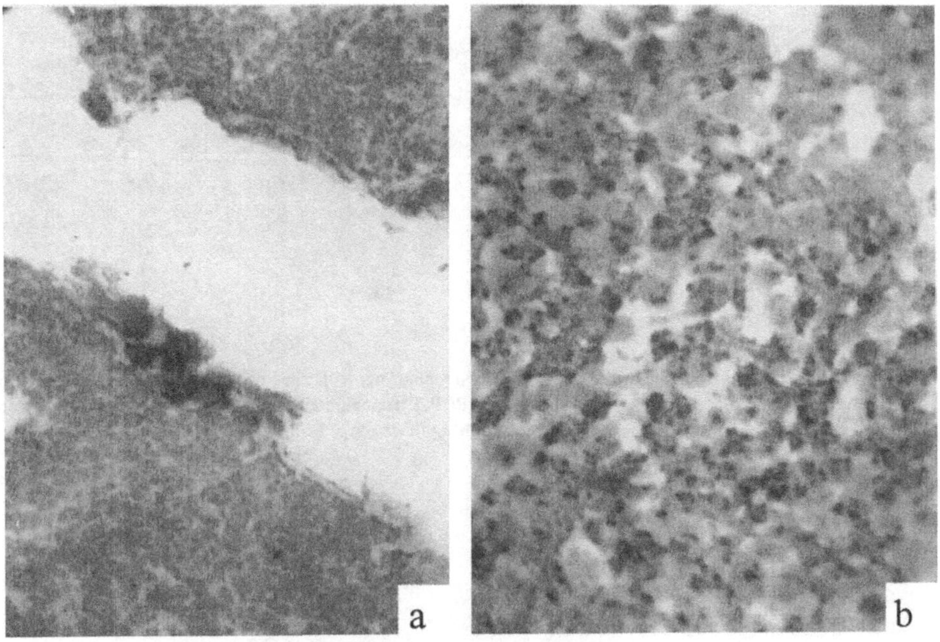

Figure 20. PRP-1 immunostaining in intact rat lymph node. (a) Accumulation of PRP-1-positive must cells on the surface of cortical sinus. Magnification x 400. (b) Dark brown PRP-1-positive granules in the germinal center of follicle. Magnification x 1000.

In the germinal centers of follicle between developing B cells and accessory cells, numerous dark-stained spherical granules are visible in a high power microscope in the intracellular substance and cell surfaces (Fig. 20b).

In the sections with toluidine blue staining, the majority of the granules show metachromatic reaction, which are perhaps the granules of degranulated mast cells[109].

The results of immunohistochemical studies with the use of monoclonal antibodies against PRP-1 demonstrated the peptide localization both in the thymus and lymphatic nodes. Defined PRP-1-containing granules were found in great quantity in B-cell zone of white rat lymph node follicles. The PRP was also discovered in great quantity in mast cells of lymph node sinus and in macrophages.

As shown, PRP-1 was localized in the cells of rat thymus vessel wall, as well as in Hassell's corpuscles. Earlier we succeeded to show that PRP-1 stimulates cell differentiation, and that its presence in thymus corroborates this function (See Chapter 5).

Thus, for the first time not only interleukins, but also novel neurohormones - PRPs with 10-15 amino acid residues along with vasopressin and oxytocin were isolated from the neurosecretory granules of the hypothalamus and neurohypophysis and identified by biological and chemical methods. The processing of these hormones from the NVAG molecules takes place during the neurosecretory granule transport from NPV and NSO cells (axonal transport) into the neurophypophysis. If we take into consideration their release into the general circulation (PRP-1 can be detected in lymphatic nodes, mast cells etc.), these compounds can be classified as hypothalamic neurohormones. PRP-1, unlike all known brain peptides, vasopressin and oxytocin, is a unique compound, possessing properties of cytokines with wide spectrum of biological activity on the thymocytes, immunocompetent cells, etc.

2.4. INTERACTIONS OF PRP-1 WITH COAGULATION CASCADE COMPONENTS : A POTENTIAL ROLE FOR PRP-1 IN THE METABOLISM OF FACTOR Xa/ANTITHROMBIN III COMPLEX.

It was shown that PRP-1 prevents erythrocyte hemolysis, inhibits aggregation of thrombocytes and accumulates in lymphocyte and erythrocyte membranes. With the help of monoclonal antibodies, it was possible to establish that PRP-1 is released from the neurohypophysis into the general blood flow and is accumulated to a great amount in lymphocyte and erythrocyte membranes[31,32]. Because its effect on thrombocytes, studies on PRP-1 interaction with one of the coagulation cascade enzymes, factor Xa, were initiated. In cooperation with Prof. Brecher and coworkers, we studied the effect of PRP-1 on the interaction between coagulation factor Xa (FXa) and the primary naturally occurring inhibitor of FXa, antithrombin III (ATIII), by SDS-PAGE. PRP-1 appears to have a biphasic affect on the FXa-ATIII complex. High levels of PRP-1 (8 μg) appear to promote the transformation of FXaα (~ 52 kDa) to FXaβ (~ 47 kDa). They further promote formation of a tertiary FXaβ-ATIII (~ 62 kDa) complex from a primary FXaβ-ATIII (~104 kDa) complex, while also reducing formation of the 3° FXaα-ATIII (~ 66 kDa) from its presumed precursor, the 1° FXaα-ATIII complex (~ 109 kDa).

In contrast, lower level of PRP-1 (4μg) appears to depress the formation of FXaβ from FXaα, and also depresses the transformation of primary FXaα-ATIII and FXaβ-ATIII complexes into the tertiary complexes. The ratio of β:α FXa in the 1° complex is approximately 1:1 and approaches 2:1 in the 3° complex. A presumed inactive γ FXa (~ 35 kDa) is also observed in mixtures of FXa/ATIII ± PRP-1. 8 μg of PRP-1 stimulate the appearance of larger quantities of the γ form when pre-added with FXa, ATIII, or a

FXa/ATIII mixture. At 1 and 4 μg of PRP-1, there is a statistically significant decrease in the γ species. It is suggested that the γ species may be generated by hydrolysis of the complexes by free FXa. Both high and low levels of PRP-1 promote the appearance of a modified form of ATIII (ATIIIM; apparent MW ~ 58 kDa). In the presence of PRP-1, a statistically significant increase in ATIIIM is observed in nearly all cases. Addition of 8μg of PRP-1 resulted a 20% increase, whereas addition of 4 μg of PRP-1 to FXa resulted in an increase by 16.7%, while nearly simultaneous addition of FXa, ATIII, and 4μg of PRP-1 yielded an 18% increase in ATIIIM. Preincubation of 4μg of PRP-1 with ATIII did not cause such increase in ATIIIM, and preincubation of 1 μg of PRP-1 with ATIII exhibited a 10% decrease in ATIIIM formation. While the mechanism of the action of PRP-1 in ATIIIM formation is not yet understood, it is of interest to speculate that the neuropeptide might bind FXa and accelerate its proteolytic activity toward the FXa-ATIII complex, or alternatively, bind the FXa-ATIII complex altering its conformation sufficiently to accelerate binding/cleavage by free FXa. This may be quite analogous to the potential of heparin to bind thrombin thereby affecting catalytic reactions.

Data obtained indicate that PRP-1 in pico- and nanomolar concentrations produces a unique specificity in the Xa coagulation factor (Coggin M.H., Srapionyan R.M., Galoyan A.A., Brecher A.S., 2002, J. Neurochemistry, v. 81 p.7).

PRP-1 appears to have a biphasic effect on the complex between FXa and ATIII. Increase in complex formation, conversion of free FXaα to free FXaβ, and conversion of 1° FXaβ-ATIII to 3° FXaβ-ATIII take place at high PRP-1 concentration. At low PRP-1 concentration, a reduction in total complexes, a reduction in conversion of free FXaα to free FXaβ, and a reduction in conversion of 1° FXa-ATIII to 3° FXa-ATIII occur.

While the mode of action of PRP-1 upon various components participating in FXa/ATIII interactions has yet to be established, PRP-1 has been demonstrated to affect molecular mechanisms of blood clotting.

3

INTERACTION OF PRP WITH CYTOKINES, HORMONES, AND NEUROTRANSMITTERS

3.1. EFFECT OF PRP-1 AND LYSATES OF NEUROSECRETORY GRANULES ON THE EXPRESSION OF CYTOKINES IN MACROPHAGES AND ASTROCYTES.

When comparing the effects of granule lysates (NGH and NGN) on L_{929} fibroblasts with the effects of neuropeptides on these cells, according to the data obtained, PRP-1 alone had almost no effect on the fibroblasts. In the presence of PRP-1, TNF-α activity was 1.2 U/ml (Fig. 21 A2). At the same time, in the NGH lysate, TNF-α activity was 42.4 U/ml, Fig. 21 A1). In the presence of vasopressin (10 μM), TNF-α activity was 34.2 U/ml (Fig. 21 A4). The combination of NGH lysate and vasopressin or oxytocin slightly increased the NGH lysate activity (Fig. 21 A5 and A6). Monoclonal antibodies (mAb) to TNF-α abolished all the effects of both NGH and polypeptides (Fig. 21B).

Next the effect of NGN lysate on TNF-α expression by LPS-stimulated macrophages was studied (Fig. 22). The experiments revealed some activity (3.1 U/ml) without stimulation of macrophages by LPS. Under LPS, TNF-α activity increased to 16.2 U/ml. Addition of different quantities of NGN lysate to the culture medium caused an increase in TNF-α activity up to 50% in the presence of LPS (Fig. 22).

Calculations showed that the content of vasopressin and oxytocin in the neurosecretory granules used in the experiments was so low that they could not produce such an effect. It follows from the above results that the neurosecretory granules themselves contained TNF-α.

Next effects of the NGH and NGN lysates as well as of PRP-1, vasopressin, and oxytocin on the ability of astrocytes to secrete inflammatory cytokines TNF-α, IL-1, and IL-6 was studied.

Our experiments demonstrated that the activity of TNF-α produced by astrocytes under LPS stimulation in the presence of the NGH lysate, vasopressin, or oxytocin was

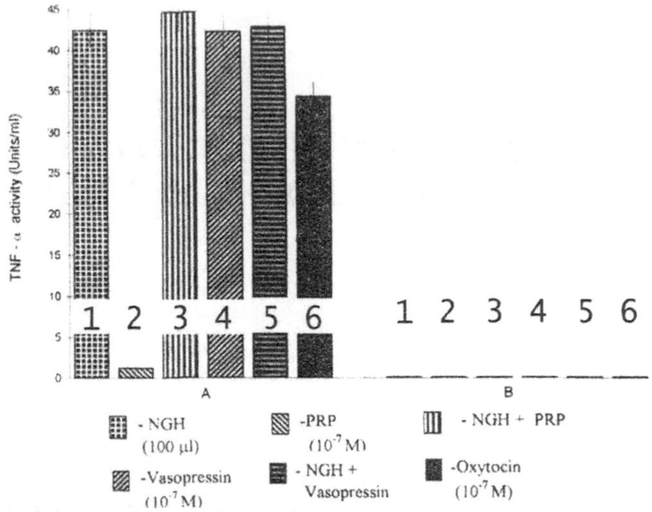

Figure 21. Effect of NGH and neuropeptides on TNF-α activity of the NGH lysate.

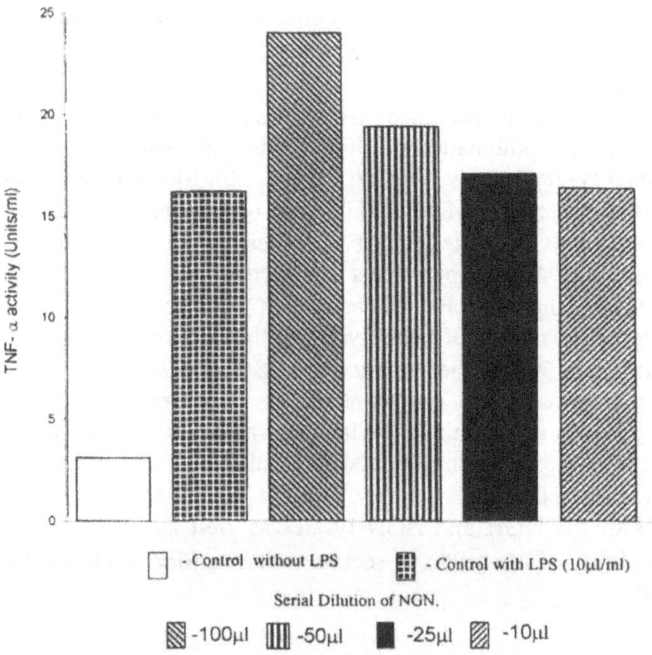

Figure 22. Effect of NGN lysate on the expression of TNF-α by LPS-stimulated macrophages.

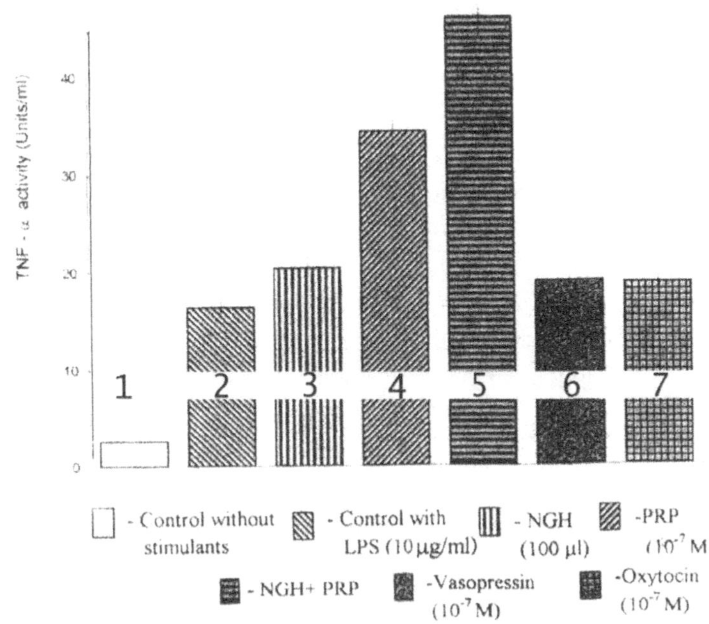

Figure 23. Effect of NGN and polypeptides on the secretion of TNF-α by LPS-stimulated astrocytes.

insignificant (Fig. 23-3, 6, 7). At the same time, PRP-1 doubled TNF-α secretion (Fig. 23-4). When PRP-1 and the NGH lysate were added simultaneously, the TNF-α activity increased almost 3 times (Fig. 23-5). So, it is possible that PRP-1 binding with some components of NGH may lead to an increase in the stimulation of TNF-α secretion by astrocytes.

It is evident from Fig. 24 A 1 that IL-1 activity (86.4 U/ml) in astrocyte culture appeared upon LPS stimulation. Under the effect of the NGH lysate, the activity achieved 108.8 U/ml (Fig. 24 A2). PRP-1 caused an increase in IL-1 activity up to 141.7 U/ml (Fig 24 A3), and maximal IL-1 activity (159 U/ml) was achieved at simultaneous addition of PRP-1 and NGH lysate (Fig. 24 A4). Meanwhile, vasopressin and oxytocin had only a slight effect on IL-1 activity (Fig. 24 A5 and A6). Injection of rabbit antiserum to IL-1 abolished all the effects of peptides (Fig. 24 B.)

Data on production of IL-6 by astrocytes were similar to those for IL-1 (Fig. 25).

We were also interested in studying effects of the NGH and NGN lysates, as well as of neuropeptides, on the antigen presentation function of astrocytes and macrophages which was determined by measuring macrophage-dependent T cell proliferation[57]. In our experiments neither the NGH and NGN lysates nor PRP-1, vasopressin, and oxytocin, stimulated antigen presentation function of astrocytes (Fig. 26A). However, in the case of macrophages, both NGH lysate and PRP-1, and especially their combination, strongly stimulated this function (Fig. 26B). The value of [^{3}H]-thymidine uptake in control (without stimulator) was 13.3 x 10^3 cpm. Under the effect of the NGH lysate it was 18.7 x 10^3 cpm (stimulation coefficient is 1.4).

Under PRP-1 effect the [^{3}H]-thymidine uptake reached 46.0 x 10^3 cpm (stimulation coefficient is 3.46). Under the combined effect of NGH lysate and PRP-1, the uptake

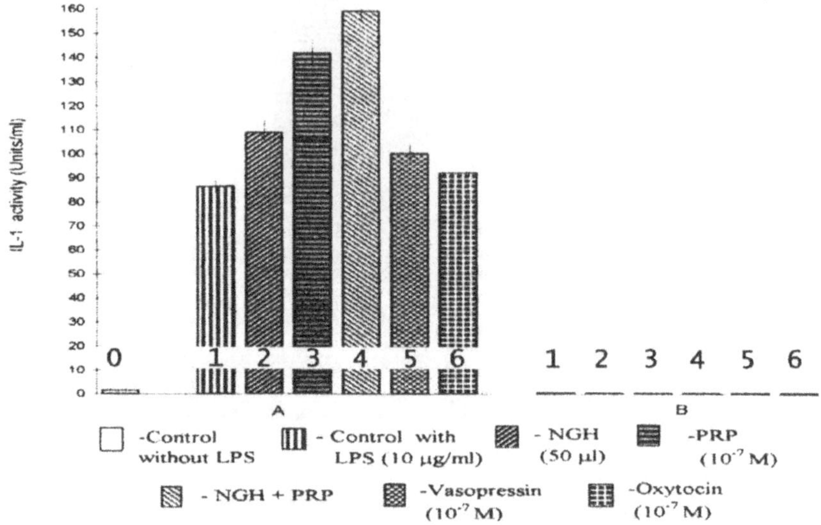

Figure 24. Effect of NGH and polypeptides on the production of IL-1 by LPS-stimulated astrocytes.

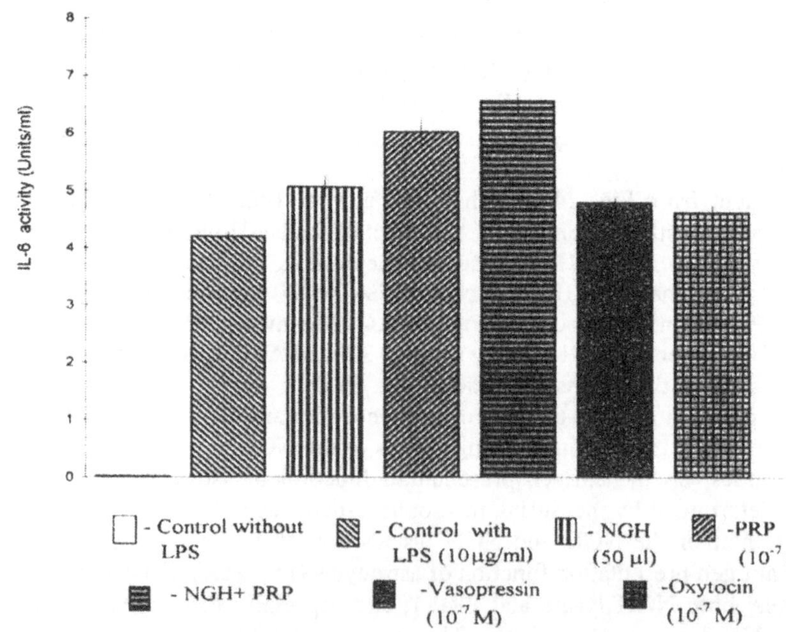

Figure 25. Effect of the NGH and polypeptides on IL-6 expression by LPS-stimulated astrocytes.

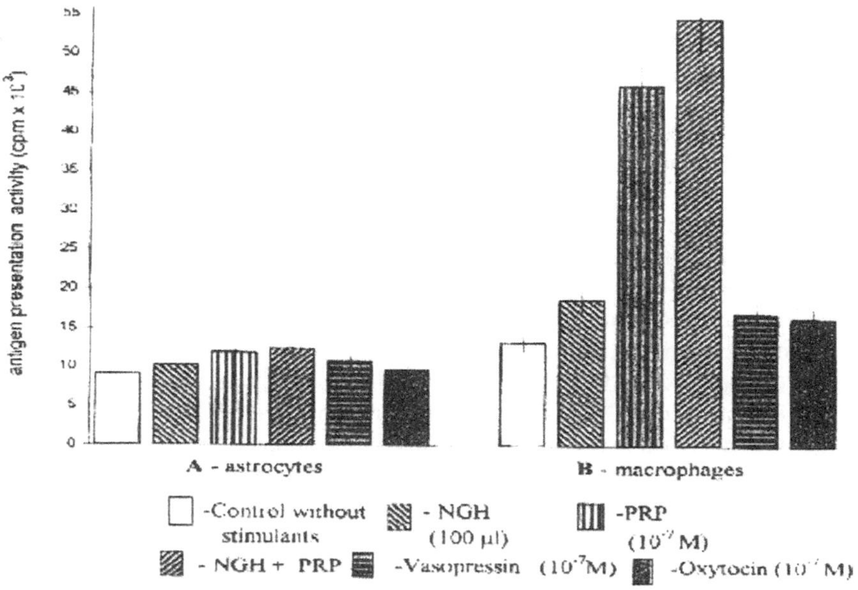

Figure 26. Effect of NGH and polypeptides on the antigen-presentation function of astrocytes and peritoneal macrophges.

value of [^{3}H]-thymidine amounted to 54.6 x 10^3 cpm (stimulation coefficient is 4.1). Vasopressin and oxytocin had no effect on the process[35,36].

Data on ILs (IL-1, IL-6) and TNF-α expression in astrocytes present evidence in favor of tight neuron-glial connection which is realized through intermediates of the brain neuroendocrine immune system. However, the action of PRP-1 on brain prolactin expression during aluminum neurotoxicosis (when prolactin content is **increased**~ 15 fold) demonstrates that PRP-1 exerts a multiple effect on the gene level, possibly by being the transcription factor for a number of genes.

3.2. EFFECT OF PRP-1 ON THE EXPRESSION OF INTERLEUKINS AND PROLACTIN IN DIFFERENT ORGANS OF MICE DURING ALUMINUM NEUROTOXICOSIS

The discovery of novel cytokines, produced by the neurosectory cells of hypothalamus (NSO and NPV)[1], as well as the synthesis of well-known cytokines in these neurosecretory cells with their following secretion to the neurohypophysis[35], raises the problem of elucidating the role of these cytokines in providing the neurophysiological and neuroendocrine function of these cells. It is known that cytokines produced in response to various factors form integrated structural-functional system (cytokine network), based on their strong mutual relations. Within this network the exhibition of the biological effect of separated cytokines is under coordinated regulation.

In this connection the most important principles of the functioning of the cytokine network appear to be their inducibility, localization of action, interrelationship, and interdependence. It is, therefore, of great interest to evaluate the influence of proline-rich polypeptide 1 (PRP-1) that also possesses the immunotropic and neuroprotective properties, on the content of IL-1, -2, and –6 in intact brain, hyppocampus, thymus, and liver in different pathological conditions and particularly under neurotoxicosis induced by aluminum.

Also it seems interesting to study the shifts in concentrations of these ILs and prolactin under aluminum toxicosis as an "experimental model of Alzheimer's disease"[111].

Alzheimer's disease (AD) experimental model was created in mice by a single subcutaneous injection of 0.20 ml of 3% solution of $AlCl_3$. For the experiments 80 white male mice weighing 30-40 g were used. The animals were divided into 3 groups: the control group consisted of intact animals; the 1st experimental group was injected with $AlCl_3$ only; the 2nd experimental group was i.p. injected with PRP-1 (15 or 1.5 μg/40 g of animal mass) on the 4th day after injection of $AlCl_3$. The animals were killed on the 8th day after the injection of $AlCl_3$. Blood serum and supernatants obtained from mice internal organs (intact brain tissue, cerebral cortex, hippocampus, thymus, and liver) were analyzed by ELISA for determination of IL-1, IL-2, IL-6, prolactin, and IGF-1. The analysis of cytokines was performed with DRG International Inc. (USA) kits, and prolactin was determined with Microwell™ FSH EIA (USA) kits. Cytokine concentration was expressed in pg/ml, and that of prolactin, in ng/ml (Table 5)[111].

Thus, it can be suggested that PRP-1 possesses modulatory action on hormone production in the anterior hypophysis, directly and/or indirectly stimulating the synthesis of prolactin. It is of considerable interest to elucidate the importance of such a high concentration of prolactin in the brain under aluminum toxicosis. Probably, prolactin plays a role of trophic factor in the brain. The role of IL-6 in hippocampus has to be elucidated. Hence, PRP-1 appears to be a powerful immunomodulator in thymus and other organs. Taking into consideration the absence of prolactin in the hippocampus, the site of hormone accumulation can be ascertained. A very strong stimulating effect of PRP-1 on the biosynthesis of prolactin in the brain was discovered (Table 5).

Results of experiments on the background of aluminum neurotoxicosis indicate that PRP-1 can be one of factors influencing the expression of prolactin in brain neurons and glial cells. ELISA results showed that in control animal serum IL-1 content was only 2.4 ± 0.7 pg/ml. After subcutaneous injection of 0.2 ml of a 3% solution of $AlCl_3$, IL-1 content increased 24 fold. Interestingly, in the case of combination of $AlCl_3$ with PRP-1 (15 μg) IL-1 content almost doubled, while with $AlCl_3$ alone (1.5 μg) its content did not change. A totally different picture was observed with IL-2 in blood serum. In comparison with control, IL-2 content in the serum under $AlCl_3$ increased by half, while combination of $AlCl_3$ with 15 μg of PRP-1 resulted in 2.5 fold increase in IL-2 content.

Under the action of 1.5 μg of PRP-1, IL-2 doubled in comparison with control and increased almost 4-fold in comparison with $AlCl_3$ experiment. IL-6 content in blood serum noticeably decreased both under $AlCl_3$ action and under combined action of PRP-1 (15 μg) and $AlCl_3$. However, PRP-1 at 1.5 μg/ml somehow increased IL-6 content (compared to $AlCl_3$), which was still much lower than control level (Table 5). Prolactin content underwent similar changes; however, in the case of 1.5-μg PRP-1 injection, prolactin content increased in comparison with control, while under $AlCl_3$ even in combination with 15 μg of PRP-1 prolactin level decreased almost 4 times.

Table 5. The content of cytokines and prolactin in serum and organs of mice during aluminum neurotoxicosis after administration of PRP-1.

Subject of study	Cytokines (pg/ml)	Control	$AlCl_3$	$AlCl_3$ + PRP (15 μg)	$AlCl_3$ + PRP (1,5 μg)
Serum	IL-1	2,4±0,7	59.64.8	129.5±7.47	62.8±6.3
	IL-2	232.0±17.6	109.6±17.6 (p<0.01)	286.4±105 (p<0.05)	430.0±20.9 (p<0.05)
	IL-6	1661.0±109.9	326.9±134.0 (p<0.01)	302.1±17.6 (p<0.01)	580.9±27.1 (p<0.01)
	Prolactine (ng/ml)	5.3±1.2	1.4±0.7 (p<0.01)	1.9±0.7 (p<0.01)	6.9±1.2 (p<0.01)
Thymus	IL-1	131,2±14,1	Traces	97.8±7.6	62.8±6.3
	IL-2	6543.0±182.3	359.3±14.9 (p<0.01)	1120.6±115.2 (p<0.05)	7200.0±234.6 (p<0.05)
	IL-6	1013.0±51.0	422.3±16.8 (p<0.01)	488.1±13.6 (p<0.01)	302.9±19.5 (p<0.01)
	Prolactine (ng/ml)	0.1±0.03	Traces	Traces	2.6±0.8L (p<0.01)
Liver	IL-1	146,0±14,9	84.9±7.0 (p<0.01)	197.3±11.4 (p<0.01)	126.4±14.3 (p<0.01)
	IL-2	399.0±21.05	120.6±13.5 (p<0.01)	105.4±12.7 (p<0.01)	248.7±14.5 (p<0.01)
	IL-6	415.9±33.0	103.6±11.5 (p<0.01)	96.1±18.2 (p<0.01)	366.7±22.8 (p<0.01)
	Prolactine (ng/ml)	2.5±0.5	-	-	-
Brain	IL-1	-	176.4±14.1	88.3±13.6	130.3
	IL-2	1590.0±105.0	624.0±25.3 (p<0.01)	1090.0±145.7 (p<0.05)	2433.6±124.3 (p<0.05)
	IL-6	526.7±23.6	1230.0±126.9 (p<0.01)	928.1±25.5 (p<0.01)	634.3±38.0 (p<0.01)
	Prolactine (ng/ml)	19.6±3.8	2.9±0.9 (p<0.01)	48.3±7.5 (p<0.01)	23.7±7.0
Hippocampus	IL-1	-	-	-	-
	IL-2	-	-	-	-
	IL-6	203.4±18.3	435.1±29.4 (p<0.01)	309.6±31.3 (p<0.05)	284.5±13.3 (p<0.05)
	Prolactine (ng/ml)	-	-	-	-

Thus, the use of low concentrations of $AlCl_3$, as was expected, results in higher levels of IL and prolactin in the brain than in the serum. It means that in the presence of PRP-1, prolactin elevation in the brain may not be caused by the pituitary. $AlCl_3$ causes significant decrease (almost to total disappearance) in IL-1 content in the thymus, while in combination with PRP-1 (15 μg and 1.5 μg) IL-1 content significantly increases and even exceeds control level (when 1.5 μg PRP-1 is given).

IL-2 content in the thymus was determined as 6544 ng/ml; after $AlCl_3$ had been administered, IL-2 content decreased almost 10 fold. When PRP-1 was administered at a concentration of 1.5 μg, IL-2 content reached 7200 ng/ml, which was higher its level in normal state. With $AlCl_3$ at the same time a tendency of IL-6 content to decrease was observed, while PRP-1 did not abolish but, on the contrary, augmented this IL decrease in the thymus. The IL-6 content in the thymus was insignificant, and PRP-1 at a concentration of 1.5 μg, increased IL-2 content 5 times (2433.6 ng/ml) in comparison with $AlCl_3$ case – 624 ng/ml. IL-6 content significantly increased with $AlCl_3$ administration. PRP-1 at the two concentrations has a tendency of increasing the IL-6 content.

It was of special interest to observe a shift in prolactin content in the brain. In comparison with control, prolactin content decreased ~ 4 times: 19.6 ng/ml in control and 2.9 ng/ml when given with $AlCl_3$ ($p < 0.01$). PRP-1 at a concentration of 15 μg and 1.5 μg increased prolactin content 16 and 8 times, respectively. In the hippocampus it was not possible to detect either IL-1 or IL-2 content in normal state or in the case of aluminum intoxication.

IL-6 is detected only in the hippocampus; at the action of PRP-1 only an insignificant increase in IL-6 in comparison with control was observed.

Thus, PRP-1 stimulates TNF-α expression in macrophages, astrocytes, and the expression of IL-1 and IL-6 in astrocytes. During aluminum neurotoxicosis PRP-1 stimulates IL-1 and IL-2 expression in the thymus and IL-2 and prolactin in the brain. As was shown, PRP-1 stimulates growth hormone expression in BALB/c mouse fibroblasts. These data demonstrate that PRP-1 is a unique factor of IL and hormone expression in different cellular systems. These facts present evidence in favor of possible transcription of a number of genes under PRP-1 influence.

To confirm this hypothesis, it is necessary to carry out some sophisticated experiments on the gene level.

3.3. PRP-1 INCREASES THE EXPRESSION AND RELEASE OF HUMAN GROWTH HORMONE IN THE CULTURE OF BALB/c MOUSE FIBROBLASTS

Our data presented evidence in favor of neurotrophic properties of PRP-1 such as stimulation of GFAP expression and release in astrocyte culture. Those facts let us suppose that PRP-1 can enhance the expression of human growth hormone (GH) from mouse fibroblasts transfected with human GH gene (clone BALB/c-GH).

It is known that GH is important for the development and functioning of the immune system[112]. It was demonstrated that at low GH level (dwarfness and other pathologies) the inhibition of humoral and cellular chains of the immune response takes place[113]. In this respect the elucidation of the possibility of GH stimulation by hypothalamic immune modulators was of interest.

The goal of this investigation was to study the capability of PRP-1 to influence the production of human GH in a culture of transfected mouse fibroblasts.

Two-site ELISA was used for GH determination[114]. 4 x 10^4 of BALB/c-GH cells (clone 5) were placed in 24-well Costar plates (ICN, USA) with a nutritious medium (NM) RPMI 1640 (ICN), containing 10.0% fetal calf serum (ICN) in a volume of 2.0 ml, and 100 μg of canamycin (ICN). The cels were incubated at 37°C under 10.0% CO_2 for 72 h until obtaining the monolayer. Later on the cells were washed with fresh NM and reincubated under the same conditions for 24 h. 200 μl of the medium were taken out of each well to determine GH concentration. Titertek plates (ICN) were incubated with a mAb against human GH (10.0 μg/ml) at 4°C for 12 h in 0.02 M sodium-carbonate duffer at pH 9.6, 50 μl per well. After washing the unsorbed mAb with phosphate buffer solution (ICN), 50 μl of the test solution and the dilutions of a standard were added and incubated for 1 h. Then after washing 3 times, rabbit polyclonal Abs against GH were added into the wells (50 μl), and the plates were placed in the incubator for 1 h. After 3-times washing of the wells, Abs against rabbit immunoglobulines (by 50 μl) labeled with horse-radish peroxidase (Armersham, USA) were added, and the plates were incubated for 1 h. Then the content of the wells was poured out, they were washed many times, and 100 μl of a solution (10 ml of 0.1 M citric acid, 200 μl of o-phenyldiamine in ethanol (40 μl/ml) and 4.0 μl of 33% H_2O_2) were added. The plates were incubated for 20 min at 20°C, then 100 μl of 2N H_2SO_4 was added to stop the development of staining. The results were recorded spectrophotometrically at a wave length of 492 nm. For statistic analysis Student's t-test was used.

The present investigation demonstrated the ability of PRP-1 to modulate the expression of GH in the culture of mouse fibroblasts. The fibroblasts in culture reached the monolayer on the 4th day of their cultivation constantly producing GH. On the 5th day of cultivation the measurements were performed (Fig. 27, 28). PRP-1 was added into the wells at a wide range of doses (10^{-3} – 10^{-12} g/ml) (Fig. 27). Within indicated dose range PRP-1 after its single application appeared to be capable of 1.7 - 2.6 fold increase of GH in the cultures ($p < 0.001$).

It was interesting to test the ability of PRP-1 to promote GH production in the cultures of fibroblasts of various maturity. PRP-1 was used in a dose of 1 x 10^{-11} g/ml either as a single application (on the 1st, 2nd, 3rd, and 4th day), or four times application (on the same days of cultivation).

The results represented in Fig. 28 were evaluated on the 1st day after PRP-1 application in each test. Thus, PRP-1 in all cases increased the daily production of GH 1.5 - 2.74 times ($p < 0.001$).

It is interesting to note that on the 4th day (accomplishment of fibroblast monolayer formation) the effect of PRP-1 4-time application differed slightly from PRP-1 single application.

These data served a basis for elucidation of the dependence of GH expression under PRP-1 on the density of fibroblast population in culture. Cultures with various quantities of fibroblasts (1 x 10^3, 1 x 10^4, 4 x 10^4 per one ml) were used, as well as various doses of PRP-1 (1 x 10^{-5}, 1 x 10^{-8}, 4 x 10^{-11} g/ml) (Fig. 29). PRP-1 was applied once on the 4th day, and the results were evaluated on the 5th day of cultivation. Data presented in Fig. 29 presented evidence that PRP-1 exerted a strong effect on GH production independently on the quantity of fibroblasts in the culture. This effect depended only on PRP-1 dose.

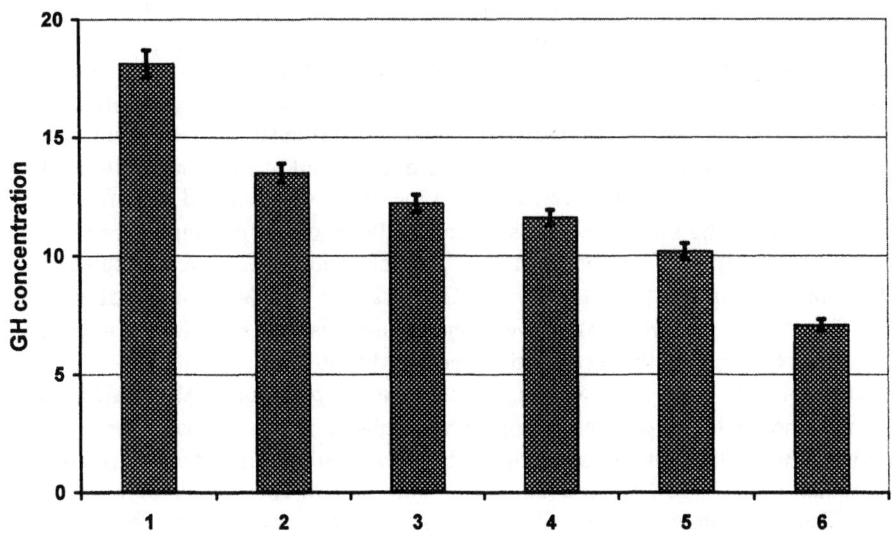

Figure 27. Changes of GH total production in mouse fibroblast cultures under PRP-1 action. Ordinate, GH concentrations (μg/ml). Abscissa, PRP-1 doses (g/ml): 1 - 10^{-12}, 2 – 10^{-11}, 3 – 10^{-10}, 4 – 10^{-9}, 5 – 10^{-8} and 6 – control (without PRP-1).

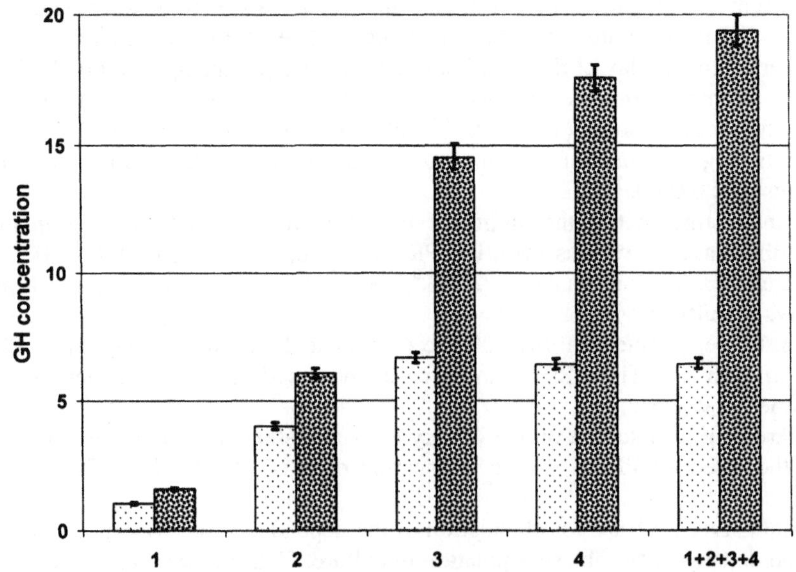

Figure 28. Changes of GH daily production in mouse fibroblast cultures under PRP-1 effect. Ordinate, GH concentrations (μg/ml). Abscissa, days of PRP-1 application (10^{-11} g/ml). White column, control without PRP-1. The black one, PRP-1 treatment.

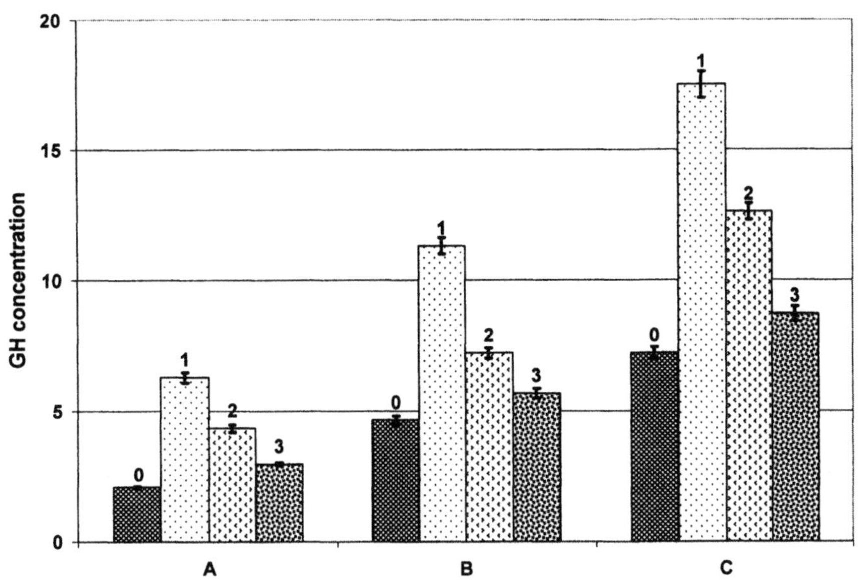

Figure 29. Dependence of GH production on the concentration of fibroblasts in culture under different PRP-1 doses. Ordinate, GH concentrations (μg/ml). Abscissa, quantity of fibroblasts per ml: A - 10^3, B – 10^4, C – 4 x 10^4; PRP-1 doses are: 0 – control without PRP-1, 1 - 10^{-11}, 2 – 10^{-8}, 3 – 10^{-5} g/ml.

For example, a PRP-1 dose of 1 x 10^{-5} g/ml increased GH production 1.35, 1.32, and 1.3 times in fibroblast cultures with 1 x 10^3, 1 x 10^4, and 4 x 10^4 cells per ml, respectively ($p < 0.001$). PRP-1 dose 1 x 10^{-8} g/ml increased GH production at the indicated concentrations of fibroblasts 1.65, 1.66, and 1.68 times, respectively ($p < 0.001$), and PRP-1 dose 1 x 10^{-11} g/ml increased GH production in the same conditions 2.56, 2.57, and 2.6 times, respectively ($p < 0.001$).

Thus, the investigations carried out revealed the ability of PRP-1 to increase human GH production in transfected mouse fibroblast cultures. PRP-1 increased the total and daily production of the hormone at the stage of monolayer formation, as well as at the stage of mature monolayer of fibroblasts. The observed effect depended on PRP-1 dose and did not depend on the time of its application and the monolayer density in culture.

The obtained data indicate that PRP-1 can participate in the expression and production of GH by adenohypophysis cells, as well as releasing factor for GH.

PRP-1 interrelation with the enzymatic system of metabolism of known neurotransmitters in the brain and lymphoid organs (thymus, spleen, etc.) are of major importance, and thus, PRP is a powerful regulator of neuronal as well as of immunocompetent cell activity.

Our joint experiments with Dr. Hambardzumyan showed that PRP-1 injection into rats in a dose of 0.2 mg/kg decreases brain glutamine content, causing statistically significant suppression of phosphate-activated glutaminase activity and decrease in GABA content by 35 – 40%. Simultaneously, in lymphocytes and splenocytes, PRP-1 increases phosphate-dependent glutaminase activity 1.5 – 2 times. Experiments in this direction are in progress. However, our data show, that the glutamine-glutamate GABA system directly joins PRP-1 immunomodulatory action, which in turn leads to the preservation of glutamine by inhibition of glutamine utilization enzymes in the brain along with sharp activation of phosphate-dependent glutaminase in lymphoid organs. These results provide evidence in favor of glutamine-glutamic acid-GABA participation in the mechanisms of PRP-1 action during T-lymphocyte differentiation and splenocyte activation, in other words in the activation of the immune system itself.

4

ANTIBACTERIAL AND ANTIVIRAL ACTIVITY OF PRP-1 AND SOME IMMUNOLOGICAL MECHANISMS OF ITS ACTION

In our recent publications immunological methods of investigations of antibacterial properties of PRPs were described[115,116] as well as some fundamental biochemical mechanisms of their antibacterial action[31,117,118]. Data obtained indicate that PRP-1 is a strong atibacterial. Antibacterial properties of PRP-1 were tested on the following species of bacteria: *Salmonella typhimurium*, *Salmonella cholerae suis*, *Salmonella typhi*, *Escherichia coli*, *Pseudomonas aeruginosa*, *Shigella Flexneri*, *Shigella Sonnei*, *Staphylococcus aureus*, *Streptococcus pneumoniae.*

All these highly virulent strains of bacteria were received from I. Metchnikoff Vaccines and Sera Institute (Moscow, Russia) and displayed typical cultural, biochemical, and morphological properties. Before use, microorganisms were cultured in 100 ml of 2.5% Nutrient Broth (Difco Lab., Detroit, MI) for about 18 h at 37°C. Then the bacteria were collected by centrifugation for 10 min at 1,500 g, washed twice with PBS, and resuspended in gelatin-Hanks-balanced salt solution (HBSS) at a concentration of 1 - 2 x 10^8 bacteria/ml. 500 µl of each sample dilution was injected i/p into mice. Portions of suitable dilutions were also inoculated onto agar plates for precise enumeration of bacterial cells (colony-forming unit, CFU). Identification of CFU of microorganisms was observed with different type-specific agglutination standard sera. Female pathogen–free BALB/c, C57BL/6J, C3H/HeJ, (CBAxC57BL/6J) F_1 strains of mice at the age of 6-8 weeks (Laboratory of Animals "Stolbovaya" of the RAMS, Moscow) were used in the experiments[115-119].

Determination of lethal doses (LD) of bacteria in mice. Bacteria of each strain were injected i.p into 10 mice in doses ranging widely in CFU. The surviving mice were counted daily for 21 days, and the 50% and 100% LDs (LD_{50} and LD_{100}) were calculated counting 21-day survivors using Probit analysis[120].

Growth of bacteria in the internal organs of infected mice. At different time points after bacterial challenge mice were killed by cervical dislocation, and their peripheral blood, liver, spleen, intestinum duodenum, mesenterial lymphatic nodes, intestinum jejunum, and intestinum crassum were obtained aseptically. 100 µl of each tissue homogenate dilution was spread over nutrient agar broth. Colonies of bacteria were counted after 18 h of incubation at 37°C.

Determination of antibacterial antibody formation in blood of infected mice. In blood sera of mice infected with *S. typhimurium* and *S. cholerae suis* the titers of anti-O-antibodies to *Salmonella* O-antigens were estimated by passive hemagglutination test with erythrocyte ultrasound-O-diagnostic system[116].

Determination of bactericidal activity of macrophages (MF). The method of direct measurement of the bactericidal activity of peritoneal macrophages (PMF) was used with minor modifications[117].

Briefly, 1.0 ml of a suspension of viable *S. typhimurium* in sublethal dose in 0.1% gelatin-Hanks solution with 10% newborn calf serum was injected i/p into mice. After 3 min the mice were killed, and peritoneal cells were collected. To remove the extracellular bacteria, the cell suspension was washed three times with ice-cold gelatin-Hanks solution and centrifuged for 4 min at 110 g; the MF concentration was then adjusted to 6×10^6 cells/ml. MF containing bacteria were reincubated at 37°C, and at various time points the number of viable intracellular bacteria was determined by a microbiological assay. PRP-1 was administered 1 h before the challenge of mice with bacteria.

The direct action of PRP-1 on the growth of bacteria in *in vitro* cultures was estimated microbiologically. It was established that PRP-1 did not possess an etiotropic effect. PRP-1 did not change quantitative and qualitative parameters of bacterial growth.

4.1. Effects of PRP-1 on the survival of mice infected with different strains of gram-negative and gram-positive bacteria at lethal doses

It was found that PRP-1 administered into mice as a single bolus dose (10^{-8} g/mouse), 24 h before their challenge, enhanced survival of mice infected with lethal doses of bacteria 1.64 - 3.8 times ($p < 0.001$) and had a pronounced protective effect during the period of infection development (Table 6).

Table 6. Effect of PRP-1 on the change in survival rate of (CBAxC57BL/6J)F_1 mice infected with different bacteria at lethal doses.

Bacteria	Survival rate (%) of infected mice	
	Treated with PRP	Control (without PRP)
Salmonella typhimurium	78.6 ± 1.8	32.2 ± 2.2
Salmonella cholerae suis	70.0 ± 1.8	34.6 ± 1.8
Salmonella typhi	58.2 ± 1.3	26.3 ± 1.3
Escherichia coli	62.4 ± 2.1	28.6 ± 1.4
Pseudomonas aeruginosa	88.2 ± 1.9	23.2 ± 1.2
Shigella Flexneri	44.4 ± 1.4	25.6 ± 1.4
Shigella Sonnei	36.4 ± 1.5	22.2 ± 1.2
Staphylococcus aureus	43.5 ± 1.5	21.5 ± 0.5
Streptococcus pneumoniae	20.0 ± 2.0	9.2 ± 0.6

4.2. EFFECTS OF PRP-1 ON THE GROWTH OF BACTERIA IN THE INTERNAL ORGANS OF MICE INFECTED WITH *S. typhimurium*

It was established that after PRP-1 administration the duration of the persistence of the microorganisms in the body was reduced and the elimination of bacteria from the body was enhanced. Thus, the number of viable bacteria in internal organs of PRP-1-treated mice was decreased as follows: in blood by a factor of 4.02, in liver by a factor of 1.97 - 2.34, in spleen the decrease coefficient was 4.04 - 2.02, in intestinum duodenum it was 2.05 -1.57, in mesenterial lymphatic nodes, 2.32 - 2.74, in intestinum jejunum, 3.4 - 2.16, in intestinum crassum, 1.41 - 1.8 ($p < 0.001$; the data correspond to day 5 and day 10 of observation, Table 7). No viable bacteria were observed on day 10 in blood and on day 21 in other organs of PRP-1-treated mice, unlike controls.

Table 7. Effect of PRP-1 on the number of viable bacteria in the internal organs of (CBAxC57BL/6J)F_1 mice infected with *S. typhimurium*

Organs	Number of viable bacteria per organ (M ± m, lg_{10} x 4)					
	Days after infection					
	5		10		21	
	Control	PRP-treated	Control	PRP-treated	Control	PRP-treated
Blood	2.25 ± 0.04	0.56 ± 0.08	0.53 ± 0.01	0	0	0
Liver	2.58 ± 0.05	1.32 ± 0.03	2.27 ± 0.04	0.9 ± 0.0	0.69 ± 0.01	0
Spleen	2.18 ± 0.03	0.54 ± 0.01	1.03 ± 0.02	0.51 ± 0.01	0	0
Intestinum duodenum	2.26 ± 0.04	1.1 ± 0.02	0.96 ± 0.01	0.61 ± 0.01	0.67 ± 0.01	0
Mesenterial lymp. node	2.27 ± 0.04	0.99 ± 0.01	2.58 ± 0.05	0.94 ± 0.01	1.29 ± 0.02	0
Intestinum jejunum	2.28 ± 0.04	0.67 ± 0.01	1.34 ± 0.02	0.62 ± 0.01	0.64 ± 0.01	0
Intestinum crassum	2.29 ± 0.05	1.62 ± 0.03	1.66 ± 0.02	0.9 2 ± 0.01	0.98 ± 0.02	0

4.3. EFFECTS OF PRP-1 ON ANTI-MICROBIAL ANTIBODY PRODUCTION IN MICE INFECTED WITH *S. typhimurium* AND *S. cholerae suis* IN LETHAL DOSES

The results of our experiments showed that the injection of PRP-1 enhanced the production of specific antibacterial antibodies (Ab) in infected mice 1.48 - 2.2 times ($p < 0.001$). Thus, the titers of anti-O Ab on 5th, 10th, and 21th days of observation were higher in PRP-1-treated mice infected with *S. typhimurium* or *S. cholerae suis* 2.17 and 2.19, 1.48 and 1.42, 1.84 and 1.66 times, respectively ($p < 0.001$). The highest level of Ab production was observed on day 5. It was of interest that even on day 21 the levels of Ab in PRP-1-treated mice remained 1.84 and 1.66 times higher than those in controls, showing the pronounced immunostimulatory effect of PRP-1 (Table 8).

Table 8. Effect of PRP-1 on the anti-*Salmonella* antibody production in (CBAxC57BL/6J)F_1 mice

Treatment	Days after treatment	Titer of anti-O antibodies (log_2)		Stimulation coefficient
		Control	With PRP	
S.typhimurium	5	1.82 ± 0.2	3.95 ± 063	2.17
	10	1.65 ± 0.1	2.44 ± 0.16	1.48
	21	1.16 ± 0.1	2.14 ± 0.13	1.84
S.cholerae suis	5	1.64± 0.18	3.6 ± 0.3	2.19
	10	1.46± 0.12	2.08 ± 0.13	1.42
	21	1.18 ± 0.1	1.96 ± 0.1	1.66

4.4. EFFECT OF PRP-1 ON BACTERICIDAL ACTIVITY OF MACROPHAGES

It was shown that PRP-1 enhanced the bactericidal activity of peritoneal MF by means of increasing the intracellular killing of bacteria *in vitro* 1.45 – 2.5 times after phagocytosis *in vivo* ($p < 0.001$) and their clearance from the circulation. PRP-1-caused an increase in bactericidal activity of MF at 30, 60, 90, 120, and 180 min of observation which corresponded to factors of 2.5, 1.8, 1.51, 1.49, and 1.45, respectively ($p < 0.001$) (Table 9).

Table 9. Effect of PRP-1 on the rate of intracellular killing of *S. typhimurium* by peritoneal macrophages

Stimulants	Intracellular killing of bacteria (%) at various time points (M±m) (min)				
	30	60	90	120	180
Control	24.6 ± 3.1	51.3 ± 4.6	62.2 ± 5.2	66.1 ± 5.4	68.2 ± 5.6
PRP-treated	61.3 ± 4.6	92.3 ± 8.2	94.2 ± 8.4	98.4 ± 8.6	98.8 ± 8.4

4.5. EFFECT OF PRP-1 ON THE SECRETION OF INTERLEUKIN-1 BY PERITONEAL MACROPHAGES OF MICE INFECTED WITH *S. typhimurium* IN SUBLETHAL DOSES

Our studies revealed the block of IL-1 secretion by MF and the ability of PRP-1 to abolish this bacteria-induced dysfunction. PRP-1 increased IL-1 synthesis by peritoneal macrophages (PMF) in mice infected with *S. typhimurium* 3.6 times ($p < 0.001$). Bacteria injected into mice in sublethal doses caused 2.64-fold reduction in macrophage IL-1 ($p < 0.001$). PRP-1 injection stimulated IL-1 synthesis 3.6 times ($p < 0.001$); IL-1 level was 1.36 times higher then its initial level ($p < 0.001$). This observation indicates the immunomodulatory effect of PRP-1.

Observed IL-1 activity was the sum of IL-1α activity (77.3 - 79.6%) and IL-1β activity (20.4-22.7%). The specificity of the obtained results was confirmed in the presence of anti-IL-1α and anti-IL-1β antisera, which completely abolished IL-1 activity in this test-system (Table 10).

Table 10. Effect of PRP-1 on the synthesis of IL-1 by macrophages of infected mice

Stimulants	IL-1 synthesis (U)		
	Without antisera	+ anti-IL-1α	+ anti-IL-1α and anti-IL-1β
Control (inactive bact.)	40.1 ± 3.6	40.1 ± 3.6	40.1 ± 3.6
Control (viable bact.)	15.2 ± 1.2	3.4 ± 0.3	0.4 ± 0.05
PRP-treated	54.7 ± 6.3	12.4 ± 1.1	0.9 ± 0.1

4.6. EFFECT OF PRP-1 ON THE ANTIGEN-PRESENTING FUNCTION OF MACROPHAGES IN MICE INFECTED WITH *S. typhimurium*

We demonstrated by the method of MF-dependent *Salmonella*-immune T-cell proliferation, that the bacteria affect the Ag-presenting function of MF. In our experiments 1.38-2.2-fold depression of this function was observed ($p < 0.01$). However, the use of PRP-1 not only abolished this defect (the functional activity was elevated 1.8-4.2 times, $p < 0.001$), but also stimulated the manifestation of the function 1.32-1.9 times compared to the initial level ($p < 0.001$). The observed effect of PRP-1 appeared to be independent of macrophage concentration in the samples indicating the correcting ability of PRP-1 (Table 11).

Table 11. Effect of the PRP-1 on the antigen-presenting function of macrophages in infected mice

Stimulants	[^{3}H] thymidine incorporation (cpm x 10^3)			
	Number of macrophages per well			
	10^3	5 x 10^3	10^4	5 x 10^4
Control (inactive bacteria)	15.1 ± 0.7	25.6 ± 1.6	41.7 ± 2.8	51.7 ± 3.3
Control (viable bacteria)	6.8 ± 0.6	12.2 ± 0.8	25.3 ± 1.8	37.3 ± 2.4
PRP-treated	28.5 ± 2.1	41.6 ± 3.2	58.8 ± 4.1	68.3 ± 4.8

4.7. EFFECT OF PRP-1 ON THE ACCUMULATION OF MACROPHAGES IN PERITONEAL CAVITIES OF MICE INFECTED WITH *S. typhimurium*

It was established that PRP-1 enhanced the accumulation of inflammatory *Salmonella*-induced MF, but not that of resident MF. The PRP-1 effect was directly dependent on the number of bacteria. Thus, at doses of 10 and 10^2 bacteria/ml, PRP-1 stimulated MF accumulation 1.82 and 2.3 times, respectively ($p < 0.001$) (Table 12).

Table 12. Effect of PRP-1 on the accumulation of macrophages in peritoneal cavities of infected mice

Stimulants	Number of macrophages/ml (x 10^5)		
	Resident	Bacteria (10/ml)	Bacteria (10^2/ml)
Control	2.3 ± 0.1	5.6 ± 0.3	9.3 ± 0.6
PRP (10^{-8} g/mouse)	3.1 ± 0.2	10.2 ± 0.4	21.3 ± 1.3

4.8. EFFECT OF PRP-1 ON THE VIABILITY OF MACROPHAGES IN MICE INFECTED WITH *S. typhimurium*

PRP-1 increased the viability of MF, especially in infected mice. Thus, PRP-1 stimulated the viability of resident MF by a factor of 1.25 ($p < 0.001$). The level of PRP-1 effect in infected mice was directly dependent on the infecting dose of bacteria. At doses of 10 and 10^2 bacteria/ml, the viability of MF under the influence of PRP-1 increased 1.77 and 2.42 times, respectively ($p < 0.001$) (Table 13).

Table 13. Effect of PRP-1 on the viability of macrophages in infected mice.

Stimulants	Number of viable macrophages (%)		
	Resident	Bacteria (10 ml)	Bacteria (10^2ml)
Control	72.6 ± 1.3	51.4 ± 0.6	38.2 ± 0.3
PRP (10^{-8} g/mouse)	90.5 ± 1.2	91.2 ± 1.4	92.8 ± 1.4

The innate immunity system of vertebrates is based on specialized cells such as epithelial cells or phagocytes (like macrophages, neutrophils, and cytotoxic T and NK cells). These cells produce not only antibacterial peptides as effector molecules, but also a number of intracellular signaling molecules that activate other cells. The activation of genes for peptide antibiotics requires start signals, receptors, and mediators (intermediates). Nitric oxide (NO) or prostaglandins can be involved here, and also cytokines like IL-1, IL-2, IL-6, and TNF-α. With exception of a few cases, signaling pathways in vertebrates are not fully clarified, but they may in the end induce transcription factors involved in the synthesis of peptide antibiotics. NK cells and cytotoxic T cells also contain a 78-residue antibacterial peptide with a hairpin structure and three disulfide bonds.

Antibacterial peptides are widely distributed as molecular effectors of innate immunity. They share cationic charge at physiological pH and hydrophobicity, factors that facilitate peptide binding and insertion into microbial membranes.

More than 300 different gene-encoded peptide antibiotics have been found in numerous species of animals, plants, and microbes[121-124].

Mature peptide antibiotics, formed by the processing from precursors, usually are composed of 13 - 80 amino acid residues. They can be divided into 3 groups:

Group I. This group consists of linear α-helical peptides without cysteines. This family includes the cecropsins, magainin, and smaller temporins (both from frog).

Group II. To this group belong cysteine-containing peptides from skin secretion: tachyplesins and protegrins from frog. The defensins (α and β) are synthesized during hematopoiesis and stored in special granules of neutrophils (α-defensins), while β-defensins and intensins are formed in the epithelia of the airways [123];

Group III. To this group belong bactericins with the over-representation of one or two amino acids, so far mostly Pro and Arg together, but also peptides rich in Trp or Gly. The shortest in this group are the apidaecins from honeybee.

The multitude of antibacterial cationic peptides has been isolated from a wide variety of animal, plant, and bacterial species[124]. By composition and secondary structure they can be divided into four major groups: 1) cecropins and magainins which exhibit a helical structure in lipid membranes; 2) defensins; 3) bactericin; 4) the forth group includes peptides that contain a high percentage of specific amino acids such as proline, arginine-rich bovine peptides Bac 5 and Bac 7, and the porcine peptide PR-39[125-130]. These peptides have been demonstrated to be capable of the permeabilization of the bacterial inner membrane. PR-39 and PR-26 are very potent inhibitors of O_2^- generation in the cell-free assay. The finding that PR-39 and PR-26 inhibite both cell-free and whole-cell O_2^- production by neutrophils, and that the inhibition require preincubation of cells with peptides for at least 45 min before PMA stimulation, suggest that these peptides act through some intracellular target, such as the NADPH oxidase components themselves. It was found that PR-39 bind to a 47 kDa protein in both human and porcine cytosol preparations. This finding shows that PR-39 binds to 47 Phox and that the binding is specific and implies that PR-39 decreases O_2^- with the help of cytosolic component of the NADPH oxidase complex[125-130]. It was established that PR-39 blocks the interaction between p-47 ^{phox} and p22 ^{phox}.

A proline-arginine (PR)-rich antibacterial peptide, PR-39, kills bacteria by a non-pore-forming mechanism, presumably by interfering with DNA or protein synthesis[121].

Many peptides are considered to act on the lipid matrix of bacterial cell membranes, destroying its barrier property and killing the bacteria.

C. Park and coworkers demonstrated[131] that buforin II kills bacteria without cell lysis and has strong affinity for DNA and RNA, suggesting that the target of buforin II is not the cell membrane but intracellular nucleic acids.

Buforin I is an antimicrobial peptide discovered in the stomach tissue of the Asian toad Bufo gargarizans[132]. Interactions of buforin II with phospholipid bilayers in comparison with magainin-2 were studied[123]. These peptides show strong antibacterial activities in vitro against a broad spectrum of microorganisms, including fungi.

The activity of buforin II, a 21-residue cationic peptide containing the residues from Thr^{16} to Lys^{36} of buforin I, is about two times higher then that of buforin I. The residue proline-11 appears to be a key structural factor for this unique property of buforin II[133].

The authors studied the interaction of buforin II with phospholipid membranes using liposomes as a model membrane system, and the results were compared with those with magainin-2. Efficient translocation of buforin II across the membranes was shown which did not induce significant membrane permeabilization or lipid flip-flop. Buforins show much stronger antimicrobial activities against a broad spectrum of microorganisms compared to magainin-2. Buforin I was reported to have a sequence homology with the N-terminus of histone H2A[134]. Several peptides are also considered to have mechanisms of action other than membrane permeabilization[134,135]. Even for these peptides the peptide-membrane interactions are important because the peptides should first bind to and then cross the cell membrane to enter the cells. Two major classes of narural antimicrobial peptides are the cathelicidins and defensins. Cyclic defencin is detected in neutrophils and monocytes from macaque rhesus monkey.

Recently, C. Park et al. purified two novel cationic antimicrobial peptides from the skin of frog Rana nigromaculata, which were named nigrocin 1 and 2. The amino acid sequences of these peptides were determined. Nigrocin 1 and 2 produce direct effect on a broad spectrum of microorganisms. Nigrocines showed high activities against gram-negative bacteria at a minimal inhibitory concentration of 10 μg/ml. Nigrocin 2 has no proline residues, while short peptides form Ranidae have highly conserved proline residues located in the middle of the helix which are essential for antimicrobial activity[130,131].

The amphipathic feature of nigrocin 2 is expected to facilitate its interaction with microbial membranes, ultimately leading to disruption of membrane integrity.

Neutrophils represent a first line of defense against infections. These cells possess multiple antimicrobial defense mechanisms, including both oxidative and nonoxidative processes of microbe killing. Antimicrobial peptides protegrins were found in pig neutrophils.

The antimicrobial activity of calcitermin was studied *in vitro*. The amino acid sequence and mass of calcitermin are equivalent to those of 15-mer C-terminal fragment of calgranulin C.

Calgranulins, which belong to the S-100 family, also possess antimicrobial activity. Calgranulin C is a newly described member of the family produced by neutrophils and monocytes. Calcitermin, human calgranulin C, and calgranulin A do not contain proline residues[136]. Cyclic defensins were detected in neutrophils.

Indolicidin is a cationic antimicrobial peptide isolated from bovine neutrophils. The cytoplasmic membrane is the site of action of indolicidin as assayed in *E. coli* by unmasking of cytoplasmic β-galactosidase due to membrane permeabilization. It was shown that the peptide is capable of increasing the transmembrane current accross planar lipid bilayers[128]. Indolicidin possesses an activity against gram-negative and gram-positive bacteria, fungi, and protozoa.

The flies challenged with bacteria produce cecropins and several other peptide antibiotics including the proline-rich peptide drosocin (which is glycosylated), an insect defensin, and the fungicidal drosomycin. The cecropins cause rapid lysis of bacteria, and thus this property was used for estimation of their biological activity[129]. All cecropins lyse bacterial membranes but not those of eukaryotic cells or mitochondria, while melittin can lyse almost every type of cell membrane. Defensins form multimeric pore-forming complexes in membranes, thereby rendering the membrane permeable. Cecropins isolated from the cecropia moth are active against *E. coli* and several other Gram-negative species[129].

Antimicrobial peptides are amphipathic cationic molecules, frequently in α-helical or β-sheet conformation, which insert into the phospholipid bilayers of microorganisms and form lytic pores. Pore-forming activity bee melittin, moth cecropin, and Xenopus skin magainin results in the permeabilization of bacterial membranes. These peptides are antimicrobial and hemolytic agents[130]. Ranidae family-produced brevinins, renalexin, and gaegurins also exert direct antibacterial effect. Noteworthy that PRP-1 produces antibacterial activity only *in vivo* by increasing the activity of immunocompetent cells.

4.9. EFFECT OF PRP-1 ON INTERFERON-γ (IFN-γ) BIOSYNTHESIS IN HUMAN MONONUCLEAR CELLS AND THE REPLICATION OF ENCEPHALOMYOCARDITIS VIRUS IN CELL CULTURE

The above-indicated data provide evidence of the probable effect of PRP-1, a new brain cytokine discovered by us, on the synthesis of interferon-γ (IFN-γ) in human mononuclear cells (MNC) in vitro, as well as of its effect on the replication of encephalomyocarditis virus (EMV) in cell culture.

IFN synthesis and titration were studied on mononuclear cells from eight healthy persons at the age of 20-40, who had normal levels of leukocytes and lymphocytes in the blood.

MNC of peripheral blood were separated in the ficoll-verografine gradient, and after washing three times in medium 199 they were placed in penicillin flasks for culturing in suspension at 1 x 10^6 cells/ml in RPMI-1640 with 10% of bovine serum. Synthesis of IFN-γ was induced by 10 μg/ml of phytohaemagglutinin (PHA). PRP-1 was administered at a concentration of 40 μg/ml. In 72 h after the incubation in thermostat at 37 C the cells were centrifuged. The supernatant was titrated by biological method for IFN presence. Two-fold dilutions of samples in a range from 1:10 to 1:160 were put into 96-well plastic dishes with cell monolayer of Hep-2 line. In 18 h the cells received 50 tissue cytopathogenic doses (TCD_{50}) of EMV. The results were calculated after 48 h. The inverse value of maximal sample dilution inducing cell monolayer defense from the virus in more than half of the infected wells was accepted as a unit of IFN. Recombinant IFN-γ (Roferon, 3 x 10^6 IU/ml, Hoffmann-La Roche, Switzerland) was used as a reference.

Two approaches were employed to study PRP-1 antiviral properties: estimation of PRP-1 antiviral action on EMV and studying PRP-1 effect on EMV replication in Hep-2 cells. To estimate the antiviral action, two concentrations of PRP-1 were mixed with 50 TCD_{50} of EMV (equal volumes in Eagle medium with 10% of serum). After 2 h incubation at 37 C the mixtures were transferred into the wells with Hep-2 cell monolayer.

Virucidal effect of PRP-1 was evaluated by the degree of its protection of cell monolayer when complete destruction of monolayer in control wells take place[137]. Three types of experiments were performed to study PRP-1 effect on EMV replication in Hep-2 cells. (1) The peptide was added in the medium 2 h before the infection; (2) simultaneously with the virus; (3) 2 h after infection. The defence of cell monolayer in more than half of the wells was considered the manifestation of antiviral effect. We should indicate here that in suspension cultures of MNC, PRP-1 at a dose of 40 μg/ml did not affect the cell viability during 72-h cultivation. Cell density in control was (765 ± 41) x 10^3 cell/ml, whereas that in the experiment was (730 ± 35) x 10^3 cell/ml. In the flasks with PHA and PRP-1 .the greatest number of MNC was observed, (940 ± 51) x 10^3 cell/ml (the difference from PHA-treated cells was statistically significant, $p < 0.02$). At the same time, the decrease in cell density in control (without PHA) was evident (422 ± 3.1) x 10^3 cell/ml. These results show the mitogenic capacity of PRP-1, because the quantity of viable cells was almost the same under separate action of PHA and PRP-1, but their joint presence during 72 h resulted in the statistically reliable synergistic effect.

Average values of IFN-γ determination in the samples are presented in Table 14. In all cases there was no IFN induction by PRP-1 (IFN titer < 10 IU/ml). PHA induced IFN synthesis in all donor MNC cultures, average titer being 45 ± 6.2. In MHC cultures with PHA and PRP-1, IFN level was higher than that in cultures with PHA alone ($p < 0.02$). So, the neurohormone did not induce IFN synthesis in MNC but caused statistically significant elevation of its level in combination with PHA.

11. In view of abovementioned literature data on the stimulating effect of PRP-1 on T-cell functions and IL-2 production, we assert that PRP-1 stimulates TH1 cells and thus promotes more active synthesis of not only of IL-2, but also of IFN under the influence of PHA effect. Hence, PRP-1 can serve as a priming in PHA-induced IFN synthesis.

Table 14. The stimulating effect of PRP-1 on production of IFN-γ in mononuclear cells

Test conditions	Titers of IFN-γ in IU/ml	Significance (p)
PRP	< 10	
PHA	45 ± 6.2	
PHA + PRP	86 ± 9.3	< 0.02

Possible toxic effect of PRP-1 on Hep-2 cells was studied in preliminary experiments. Fife-day observations showed that cell proliferation and time required for monolayer formation were not affected by PRP-1 concentrations of 10, 20, and 40 μg/ml.

To study PRP-1 effect on EMV infection, the peptide concentrations of 20 and 40 μg/ml were used. The results of the experiments are presented in Table 15. According to these data, PRP-1 possesses virucidal properties because it neutralizes the virus cytopathic effect contacting EMV outside the cells. PRP-1 effect at both doses was almost similar. Both PRP-1 doses inhibited EMV reproduction in the case of preliminary cell treatment with the peptide.

The results of other variants of the experiment were negative as both doses of PRP-1 protected the monolayer from the destruction by EMV only in 10-15% of wells. These results showed that PRP-1 at a concentration of 40 μg/ml stimulated lymphocyte proliferation in short-living cultures of mononuclear cells from peripheral blood of healthy donors. PRP-1 did not induce the synthesis of IFN-γ *in vitro* but its application resulted in statistically significant increase in IFN production under PHA stimulation. The priming effect of PRP-1 may be explained by lymphocyte proliferation and activation of their lymphokine function. As IFN-γ and IL-2 are synthsized mainly by TH1 helper cells, it is possible to suggest that PRP-1 exerts its influence mainly on this T-helper cell subpopulation.

Table 15. Virucidal action of PRP-1 and its effect on EMV replication in Hep-2 cells

Test conditions	Test varieties	Degree of Hep-2 cells defense at PRP doses			
		20 μg/ml		40 μg/ml	
		Absolute	%	Absolute	%
PRP virucidal effect	PRP + EMV	10 / 20	50	11 / 20	55
	EMV (control)	0 / 20	0	1 / 20	5
PRP effect on EMV replication	PRP + EMV	9 / 20	45	11 / 20	55
	PRP + EMV (simultaneously)	3 / 20	15	2 / 20	10
	EMV + PRP (in 1 h)	0 / 20	0	1 / 20	5

To study antiviral effect of PRP-1, murine EMV was chosen. It is well known that encephalomyocarditis viruses infect humans and cause damage to the neurons and myocardium. The reproduction of these viruses takes place in brain neurons, and they represent a suitable model for studying antiviral activity of PRP-1.

Virucidal effect of PRP-1 is realized after its contact with EMV during two hour thermostat incubation. Though the mechanism of this effect awaits further studies, we can suggest that virus-specific cellular receptors are blocked by PRP-1, and so, the virus cannot adsorb on the cell surface and enter the cells. Regretfully, little information is available about the receptors of picornaviruses to which EMV belongs. These receptors are proteins, but they have not been purified, hence, cannot be studied by ELISA.

According to our data, PRP-1 inhibits EMV reproduction in the Hep-2 cells only if administered one our before infection. PRP-1 can affect either phase I of viral reproduction (adsorption, cell entering, and "undressing") or phase II (mRNA translation, genome replication, and viral assembly).

Our results demonstrated that further studies are necessary to elucidate the mechanisms of PRP-1 effect on IFN synthesis and viral reproduction. It is worth noticing that PRP-1 does not possess an etiotropic activity to bacterial infections studied by us. Nevertheless, that problem needs further investigation.

5

PRP REGULATION OF T-LYMPHOCYTE DEVELOPMENT AND MYELOPOIESIS

5.1. INFLUENCE OF PRP-1 ON INTERLEUKIN-2-DEPENDENT FUNCTIONS OF HUMAN LYMPHOCYTES IN CULTURE AND ON THE PROLIFERATION OF JURKAT T-CELLS

Lymphocytes were isolated from tonsils by tissue homogenization and further centrifugation of cell suspensions on Ficoll density gradient (Ficoll-paque, Uppsala, Sweden).

Cells (5×10^6 cell/ml) were cultured in RPMI-1640 (Serva), containing 2 mM L-glutamine, 1 mM sodium pyruvate, 50 μM ml gentamycin, and 10% fetal calf serum (Serva). For lymphocyte stimulation, ConA or PHA were added to the incubation medium at concentrations of 20 and 10 μg/ml, respectively. Peptide concentration in the medium was 5×10^{-5} M. After 72-hour culturing, cells were washed in PBS (pH 7.4), and cytological preparations were stained according to Romanovsky-Gimza method. Relative content of blast lymphocytes was determined in these preparations by morphological criteria (size, volume, chromatin conditions, and intensity of cell staining).

Effect of PRP-1 on IL-2-dependent T-cell proliferation was studied using of ConA-induced blasts and human recombinant IL-2 (Boehringer, Mannheim) by the method described elsewhere[138]. Cell proliferation was estimated spectrophotometrically at 570 nm wavelength by multichannel autoreader using bromide-3(4,5-dimethyl-2-tetrasolyl)-2,5 diphenyl tetrasol (Sigma). Functional activity of lymphokine-activated killers (LAK) and intensity of antibody-dependent cell cytotoxicity (ADCC) were determined by the method described by us earlier[139].

For investigation of the peptide effects on the induction of humoral immune response of cultured lymphocytes to antigens of pathogen and putative pathogenic microorganisms we used formalin-fixed strains of *Shigella Flexneri*, *Klebsiella pneumoniae*, *Yersinia enterocolitica,* and *Salmonella typhimurium*. The number of antibody-forming cells

(AFC) was determined by ELISPOT assay on nitrocellulose filters coated by corresponding antigens[140].

Investigation of PRP-1 effect on PHA or ConA-induced proliferation of tonsillar lymphocytes *in vitro* showed that the most effective action of PRP-1 was observed at concentrations between 5 x 10^{-5} and 10^{-6} M. As shown in Fig. 30, PRP-1 at a concentration of 5 x 10^{-5} M suppressed the reaction of blast transformation of ConA-treated tonsillar lymphocytes. Stimulation of lymphocyte blast transformation reaction was observed in the presence of the same concentrations of PRP-1 when recombinant IL-2 was added to the incubation medium. It should be noted that the relative amount of blast lymphocytes during ConA stimulation did not differ from that of blast cells obtained after IL-2 stimulation in the presence of 5 x 10^{-6} M PRP-1.

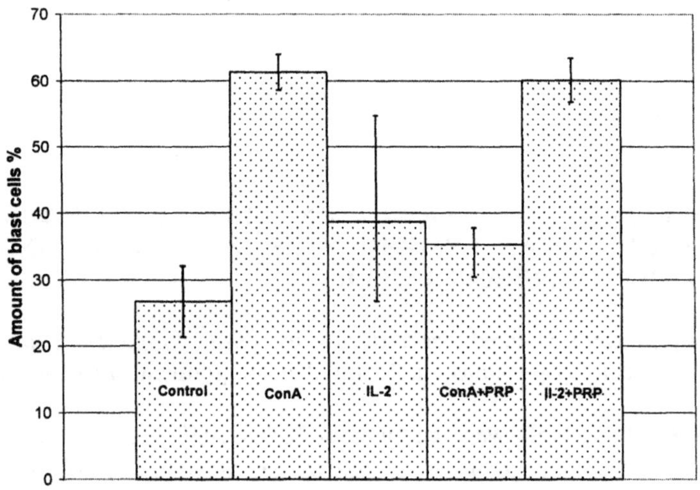

Figure 30. Effects of PRP-1, IL-2, and ConA on blast transformation of tonsillar lymphocytes in culture.

Thus, PRP-1 inhibits the growth of ConA-induced blast lymphocytes in culture when added after 72-hour culturing in the presence of ConA. PRP-1 added to the incubation medium of ConA-induced blast cells leads to significant stimulation of cell proliferation in the presence of exogenous IL-2.

The effect of PRP-1 on antibody-dependent cell cytotoxicity (ADCC) of lymphokine IL-2-activated killer cells (LAK) was studied. LAK cells were stimulated in culture by 24 hour incubation with IL-2 or ConA in the presence or absence of PRP-1. Cytotoxic effect of LAK cells and T-lymphocytes on target sheep erythrocytes or anti-erythroid serum-treated sheep erythrocytes is shown in Fig. 31.

These results show that PRP-1 significantly increased the cytotoxic activity of LAK cells during culturing of tonsillar lymphocytes in the presence of IL-2. PRP-1 stimulated the activity of ConA-induced killers towards sheep erythrocytes in the same conditions (Fig. 31). This peptide stimulated ADCC reaction in the presence of exogenous IL-2.

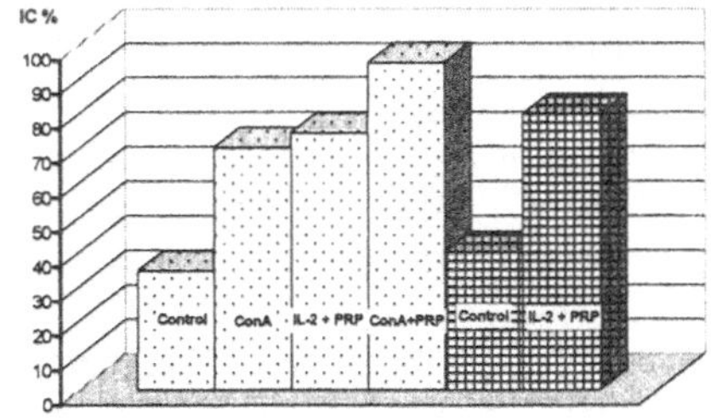

Figure 31. Stimulation of the cytotoxicity index of LAK (white) and activity of ADCC (dark) of tonsillar lymphocytes upon the action of PRP-1

We also investigated antibacterial effects of PRP-1-treated tonsillar lymphocytes using putative pathogenic bacteria, *Yersinia enterocolitica*. After 48-hour incubation of PRP-1-treated lymphocytes with live *Y. enterocolitica* and their subsequent culturing in a medium containing sucrose, mannose, lactose, urea, and phenylalanine, complete absence of bacterial sugar-metabolizing activity was observed. Induction of humoral immune response in cultured lymphocytes was stimulated in the presence of PRP-1 by formalin-fixed microorganisms: *Sh. Flexneri*, *K. pneumoniae*, *Y. enterocolitica*, *S. tiphymurium* or sheep erythrocytes. Conditioned medium of human lymphocytes activated by the mitogen from Phytolacca americana was added to the incubation medium of cultured lymphocytes. The results of these experiments showed the stimulation of antibody-forming cells producing specific antibodies to all used antigens in the presence of PRP-1 (Fig. 32).

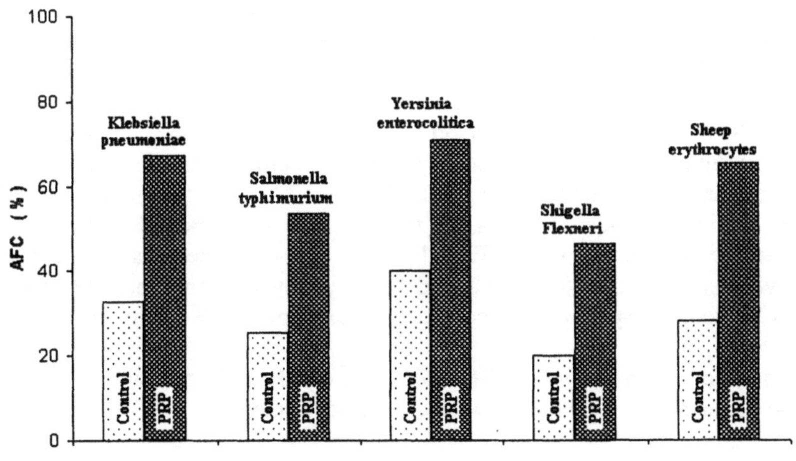

Figure 32. PRP-1 stimulation of antibody formation by bacteria-challenged B-lymphocytes.

Thus, the results of our experiments indicated that PRP-1 possesses an immunomodulatory activity, which is exposed through the modulation of IL-2-dependent cellular and humoral immune responses of tonsillar lymphocytes in culture. PRP-1 stimulated IL-2-dependent humoral (induction of antibody formation in cell culture) and cellular (IL-2-dependent proliferation of blast lymphocytes and cellular cytotoxicity) immune responses of cultured human lymphocytes.

The activation of both humoral and cellular immunity under the influence of the neurosecretory immunomodulator PRP-1 indicates the existence of neurohumoral mechanisms of integration of brain immunomodulators with immunocompetent cells, the T-lymphocytes.

We demonstrated that PRP-1 modulates PHA-mediated human lymphocyte proliferative responses in a dose-dependent manner. The highest effect was obtained at concentrations of PRP-1 in the interval $1 \times 10^{-9} - 1 \times 10^{-10}$ M.

It was shown that the degree of PRP-1 modulatory effect towards PHA-induced lymphocyte proliferation depends on functional condition of T-lymphocytes in human blood. In donors with low activity of T-lymphocytes, PRP-1 stimulates the lymphocytes by initiation of T-cell proliferation in response to PHA[141].

Thus: 1) the hypothalamic neuropeptide PRP-1 exerts significant immunomodulatory effect on human peripheral blood T-lymphocytes; 2) PRP-1 immediately acts on T-lymphocytes as shown in experiments by using of the human peripheral T-cells and human Jurkat T-cell line.

The results of *in vitro* experiments with isolated tonsillar child lymphocytes and also the effect of PRP-1 on human peripheral lymphocyte proliferative activity provide evidence that PRP-1 inhibits the blast-transformation of T-cells in the presence of PHA. PRP-1 in combination with IL-2, in contrast, increases the level of blast transformation of tonsillar T-cells, i.e., acts on IL-2-dependent processes *via* the potentiation of its influence. PHA stimulates the proliferation of peripheral lymphocytes of the general circulation.

Our experiments showed that PRP-1 has a direct effect on T-cell proliferation. This is confirmed by the inhibition of proliferation of human T-cell line Jurkat under PRP-1 influence (Fig. 33). As follows from Fig. 33, PRP-1 *in vitro* inhibits in a dose-dependent manner Jurkat cells proliferation in concentrations of $1 \times 10^{-10} - 1 \times 10^{-6}$ M[141]. In the process of T-cells activation, adaptor proteins LAT and SLP-76[142-145] play an important role. It is possible that PRP-1, bound to adaptor proteins, neutralizes their effect which leads to cell proliferation blockage. It is also possible that deregulation of cell cycle control by cycline-dependent kinases and their inhibitors occurs under PRP influence[146]. I do not preclude the possibility that PRP-1 bound to T-cell receptors deregulates tyrosine phosphorylation of their cytoplasmic domains, and by that prevents the functioning of regulatory protein phosphorylation cascade, for example, ZAP-70, mediated by Lex and Fyn[147].

It was important to study the effect of PRP-1 on the differentiation of T-lymphocytes during the neonatal and fetal period of thymus development.

5.2. REGULATION OF THYMOCYTE DIFFERENTIATION BY PRP-1 IN NEONATAL AND FETAL THYMUS

The establishment of the peripheral T-cell pool is attributed in part to cellular interactions occurring within the developing thymus[148]. The pathway of T-cell development follows a series of highly regulated steps beginning with the differentiation of bone marrow-resident pluripotent hematopoietic stem cells into T-cell progenitors that migrate to the thymus. Upon colonizing the thymic microenvironment, T-cell progenitors

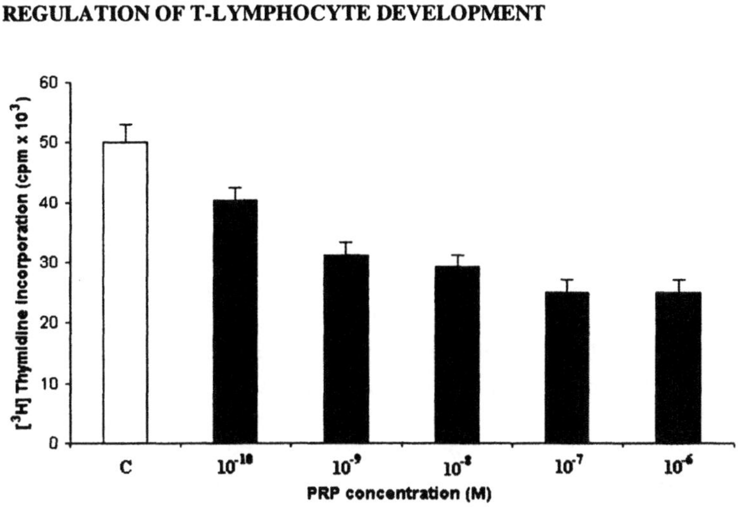

Figure 33. Effect of PRP-1 on the proliferation of Jurkat T cells.

further differentiate into mature immunocompetent T-lymphocytes through a series of stages that are characterized by the regulated expression of the coreceptor molecules CD4 and CD8. However, the acquisition of a functional T-cell antigen receptor complex TCR/CD3 is perhaps the most significant of the events that take place during early thymocyte development. Intrathymic differentiation can be divided into major developmental stages based on the expression of CD4, CD8, and TCR/CD3. T cells are finally selected on the basis of the specificities of their receptors to advance to the $CD4^+CD8^-$ or $CD4^-CD8^+$ single positive (SP) stage[149].

It is important to know what intra- and intercellular signaling mediators are involved in the regulation of early T-cell differentiation and whether brain peptides are also involved in this developmental program. These mediators, as known, are particularly important in promoting the proliferation and differentiation of cells and finally in regulating and determining the nature of the immune response[150].

Better understanding of T-cell differentiation and selection is crucial for aiding in many clinical situations, such as immunodeficiencies, autoimmune diseases, various infections, and cancer.

The aim of the present study was to determine whether PRP-1 influences the differentiation of thymocytes in fetal thymus in normal mice. All experimental procedures were performed under sterile conditions.

Mice. C57BL/6J mice were obtained from the Laboratory of Animals "Stolbovaya" RAMS (Moscow, Russia). We mixed female and male mice in a cage for a night before separation. Gestational age was calculated from mice separated (day 0) and was confirmed by size and several other developmental features of fetuses[151].

Isolation of fetuses from pregnant mice. At different ages of gestation the females were killed by cervical dislocation. Fetus-containing uteri were taken out, transferred to 100-mm plastic dishes (Millipore Corp., Bedford, MA, USA) with 30 ml of culture medium: RPMI 1640 supplemented with 10% fetal calf serum (FCS) inactivated for 30 min at 56°C, 50 μM 2-ME, 10 mM HEPES, 2 mM L-glutamine, 1x nonessential amino acid solution, 1 mM sodium pyruvate, 10 U/ml penicillin, and 100 μg/ml streptomycin. All medium components are purchased from Gibco-BRL, Gaithersburg, MD, USA, and

2-ME is purchased from Sigma Chem. (St.Louis, MO, USA). After ascertaining of gestation age the fetuses were stored on ice while preparing culture wells[152].

Isolation and organ culture of fetal thymus lobes. Fetuses on the 14th day of gestation were placed under a dissecting microscope, the abdomen turned up and the head flipped up. The chests were opened, thymus lobes were taken out and put onto a 13-mm polycarbonate membrane filter (Costar, Nucleopore Corp., USA) placed on the Helistat collagen sponges (Colla-Tec, Inc., Plainsboro, NJ, USA) in a 16-mm well of a 24-well culture plate (Costar, Cambridge, MA, USA). Five lobes were put on the membrane, and the plate was placed in an incubator with 5% CO_2[153].

Preparation of single-cell suspension from fetal thymus organ culture (FTOC). Thymus lobes were transferred into a drop containing 100 μL of a buffer (PBS, pH 7.2, supplemented with 0.2% BSA and 0.1% NaN_3) at the center of the reverse side of the lid of a 30-mm plastic dish, and the number of lobes was counted. A small piece of nylon mesh (~300 meshes/square inch, ~ 5 mm^2) was placed on the drop. Then, thymus lobes were teased gently with the 26-gauge needles, pushing them under the filter mesh. Cell suspension was transferred into a plastic tube, and the cells were counted. These cell suspensions were used for further examination of T-cell development by immunofluorescence and flow cytometry analysis[154].

Neonatal thymus organ culture (NTOC). Newborn mice thymus lobes were obtained on 0-day of the neonatal period (the 19-20th days of gestational age). The technique of NTOC preparation was identical to that of FTOC and was used for the analysis of positive selection signals inducing the generation of single-positive thymocytes[155].

Purification of immature $CD4^+CD8^+$ thymocytes. We used a simple procedure for the purification of murine $CD4^+CD8^+$ thymocytes by panning with anti-CD8 Ab[156]. Briefly, a single-cell suspension of thymocytes was added to tissue culture plates that had been coated with anti-CD8 Ab (PharMingen, San Diego, CA, USA), nonadherent cells were removed by washing, and $CD8^+$ cells were recovered.

Analysis of purified populations of thymocytes by immunofluorescent staining and flow cytometry. The cell purity was determined by standard flow cytometry method. Both unseparated (total) thymocytes and Ab-purified thymocytes were analyzed for surface expression of CD4 and CD8 by two-color analysis[157]. Unseparated or anti-CD8 panned thymocytes were stained with FITS (fluorescein isothiocyanate)-conjugated anti-CD4 mAb and biotin-conjugated anti-CD8 mAb. Leu 4 mAbs (specific for human CD3 proteins) were used as negative control. All mAbs were purchased from PharMingen. Before immunofluorescence staining and between reaction stops, the cells were washed three times in HBSS-Hanks buffered saline solution without phenol red (Gibco-BRL) containing 5% BSA (fraction Y) (Sigma). All incubations lasted 30 min, and the cells were permanently kept on ice. The percentage of FITC- or biotin-labeled cells was determined by analyzing between 3 x 10^5 and 5 x 10^5 cells. To determine the phenotype of mature cells, 10,000 events were analyzed. FACScan flow cytometer (Becton Dickinson, Sunnyvale, CA, USA) was used. The statistical analysis was performed using Minitab 10.5 Xtra software (Minitab Inc., Pennsylvania, USA) and compared with analysis of variance.

In the present study we examined the PRP-1 influence on the differentiation of mouse thymocytes in FTOC and NTOC, which represents T-cell development during fetal life and the early neonatal period[154,155]. Initially, we investigated the quantitative changes in thymocytes obtained from fresh fetal thymus lobes (*in vivo*) or from FTOC (*in vitro*). PRP-1 was injected *in vivo* or *in vitro* on the 13th day of gestation. It was established that PRP-1 enhanced the number of cells by a factor of 1.23-3.1 ($p < 0.001$) *in vivo* (Fig. 34A) and 1.34-3.5 ($p < 0.001$) *in vitro* (Fig. 34B) in fetal thymus lobes on the 13-19th days of gestation.

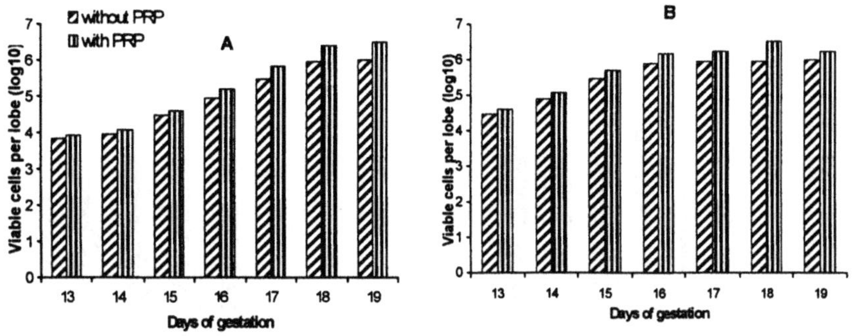

Figure 34. Counts of cells from fetal thymus lobes (*in vivo*) (A) or FTOC (*in vitro*) (B). PRP-1 was injected on the 13th day of gestation (1 μg/mice, i.v) *in vivo*, or administered *in vitro* (1 μg/1 x 10^6 cells) before the cells were counted.

It is known that most of T-lymphocytes of the immune system differentiate within the thymus along the CD4/CD8 developmental pathway by a highly ordered process termed thymic selection. The maturation status of thymocytes is usually assessed by their expression of the coreceptor proteins CD4 and CD8 and the surface density of αβ T-cell receptor (αβTCR)[158]. Three major subpopulations of T-cells exist within the thymus that exemplify the progression of thymocytes along the CD4/CD8 developmental pathway: $CD4^-CD8^-$ (double negative, DN) thymocytes which express no αβTCR; $CD4^+CD8^+$ (double positive, DP) thymocytes which express no/low αβTCR; $CD4^+CD8^-$ and $CD4^-CD8^+$ (single positive, SP) thymocytes, both of which express high surface density of αβTCR[159]. After receiving isolated single-cell suspension from FTOC, we used simple procedure for the purification of murine $CD4^+CD8^+$ thymocytes by panning with anti-CD8 Ab. PRP-1 was injected on the 14th day of gestation *in vivo* or administered *in vitro*.

Analysis of purified $CD4^+CD8^+$ thymocytes was conducted by flow cytometry. It was established that the rate of DP thymocytes in untreated mice was increased since the 14th day of gestation and peaked on the 20th day of gestation (82%, Fig. 35). Under the influence of PRP-1, the rate of DP thymocytes abruptly increased on the 16th day of gestation and peaked on the 17th day of gestation both *in vivo* (Fig. 35A) and *in vitro* (Fig. 35B) (85% and 86%, respectively). Since the 17th day of gestation and before the 20th day, the percentage of DP thymocytes decreased surprisingly (to 70%-71%) under the influence of PRP-1. Afterwards we used NTOC technique for the analysis of positive selection signals including the generation of SP thymocytes. As is known, NTOC of thymus lobes of newborn mice is a convenient model for *in vitro* investigation and stimulation of *in vivo*-generated $CD4^+CD8^+$ thymocytes[155]. PRP-1 was administered *in vivo* or *in vitro* on the 14th day of gestation (FTOC) or on the 20th (0) day (NTOC). It was established that the generation of SP thymocytes was increasing from the 17th day of gestation to the 7th day of neonatal life of mice (from 3% to 9% in $CD4^+$, and from 2% to

6% in $CD8^+$) (Fig. 36). It was interesting that PRP-1 increased the generation of both $CD4^+$ (from 7% to 21%, i.e. 1.86-3.0 times, $p < 0.001$) and $CD8^+$ (from 5% to 11%, i.e. 1.6-2.7 times, $p < 0.001$) *in vivo* (Fig. 36A) and *in vitro* (Fig. 36B). More interesting was the fact that under PRP-1 influence the number of $CD4^+$ thymocytes was higher (on average 1.44 times, $p < 0.001$) than of $CD8^+$.

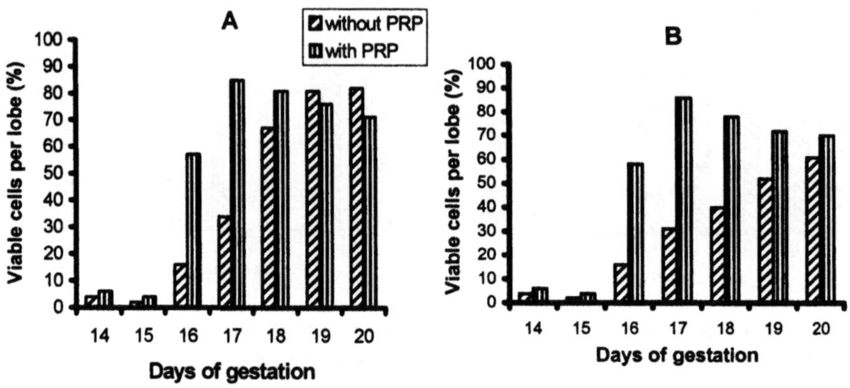

Figure 35. Immature double positive ($CD4^+CD8^+$) unseparated (total) thymocytes in fetal thymus. Thymocytes were obtained from fresh fetal thymus lobes (*in vivo*) (A) or FTOC (*in vitro*) (B). PRP-1 was injected on the 14th day of gestation *in vivo* (1 μg/mice, i/v) or administered *in vitro* (1 μg/1 x 10^6 cells) before the cells were phenotypically characterized.

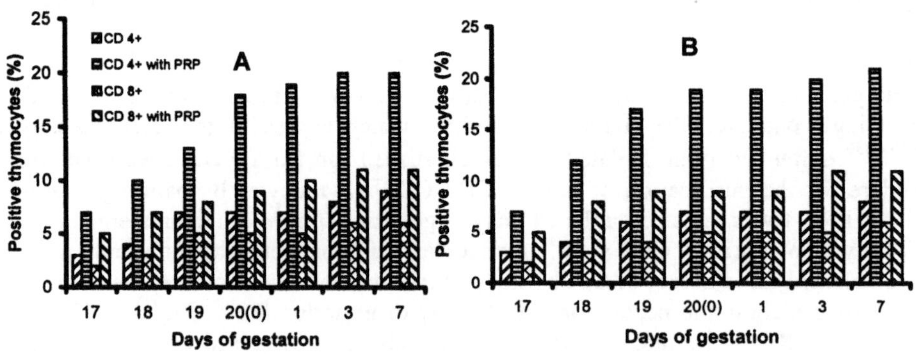

Figure 36. Mature single-positive ($CD4^+CD8^-$ and $CD4^-CD8^+$) purified populations. Thymocytes were obtained from fresh fetal thymus lobes (*in vivo*) (A) or FTOC and NTOC (*in vitro*) (B). PRP-1 was injected *in vivo* (1 μg/mice, i/v) or administered *in vitro* (1 μg/1 x 10^6 cells) on the 14th day of gestation or on the 0 (20th) day in NTOC.

Thus, our investigation *in vivo* and *in vitro* revealed that PRP-1 modulates the pathway of intrathymic T-cell development (proliferation, maturation, and surface expression of MHC) during fetal life. PRP-1 increases the total number of thymocytes in fetal thymus lobes (Fig. 34) and the rate of their differentiation (both in unseparated, Fig. 35, and purified, Fig. 36, populations).

Of special interest are two fundamental findings concerning PRP-1 action. First, simultaneously with decrease in the level of immature DP thymocytes (after the 17th day of gestation, Fig. 35), an increase in the generation of mature SP thymocytes was observed (Fig. 36). Second, the generation of CD4$^+$-thymocytes was higher than that of CD8$^+$-thymocytes (Fig. 36). It may result in subsequent reinforcement of T-lymphocyte functions.

It is well known that differentiation of T-cell precursors to mature T-cells is dependent upon both direct contact with thymic stromal cells and interaction with cytokines, and possibly, with other endogeneous bioactive substances. Of the cell surface molecules involved in the direct contact with thymic stromal cells, the TCR plays a pivotal role in this process of determination of positive selection or death[160].

Cytokines (and probably other active endogeneous molecules) were long believed to provide proliferation signals for thymocytes. This belief was based on the observations that many cytokines induce proliferation of thymocytes *in vitro*[161]. However, studies employing targeted gene knockout in mice have shown that of all cytokines which are thought to control T-cell development only IL-7, thymic stroma-derived lymphopoietin (TSLP), SCF-stem-cell factor (c-kit ligand), Fas-L and, CD30-L (both are members of the TNF receptor family) play major roles in the development of T-cells in the thymus[162,163].

However, lymphopoiesis depends probably on a cocktail of microenvironmental stimuli, exhibited to developing lymphocytes which are not yet completely defined[162,164]. At present many questions in this field are not answered. For example, what signals are necessary for maturation of the lymphoid precursors of T, B, or NK cells? And how the direction of selective expression of CD4$^+$ or CD8$^+$ is controlled? Solving of these biological puzzles will undoubtedly contribute to our understanding of the basis of human diseases like lymphoid cancers, autoimmunity, and immunodeficiency. The discovery of new molecules involved in thymocyte differentiation will definitely contribute to our understanding of this complex process.

In the thymus, the cytokines and other substances may present which initially attract precursor cells, control their slow passage through the organ, and finally control their departure. Molecules may exist which induce cell growth, cell positive selection or death, differentiation into different lineages, such as $\alpha\beta$ *vs.* $\gamma\delta$, CD4 *vs.* CD8, etc.[165]. Besides, both intrathymic and extrathymic T-cell development may need different cytokines and other regulator molecules altogether. And one of them may be PRP-1.

The thymus can produce B-cells, and the bone marrow can generate T-cells (but not myeloid cells[166]). The fact that thymus physiologically produces B- as well as T-cells suggests that the putative common leukocyte progenitors (CLP) may be candidates for the role of cells that migrate into the thymus from the bone marrow.

IL-7, Ra-dependent signaling as well as the IL-2 receptor γ-chain (γ_c) are important components for T-cell maturation[160] and intrinsic programs of differentiation[167]. T-cell maturation involves rearrangements of germinal TCR genes and the expression of various membrane markers.

Taking into account the antibacterial properties of PRP-1, the generation of mature single-positive thymocytes allows me to conclude that PRP-1 is a stimulator of the maturation of TH1 and TH2 forms. The CLP can give rise to T-, B-, and NK-cells, but not to myeloid cells. The existence of CLP also suggests the presence of common myeloid progenitors (CMP), which could be a complementary population to CLP at a site of early hematopoiesis.

The common myeloid progenitor cells differentiate into the monocytes, macrophages, megakaryocytes, eosinophiles, neutrophiles, basophiles, and mast cells.

Transcription factors regulating the development of the hematopoietic system can be functionally classified into two categories. One group, e.g. GATA-2, PU.1, and Ikaros, is required for the development of multiple lineages, whereas the other group, e.g., EBF, Pax-5, and GATA-3 functions to orchestrate the development of a particular lineage[168].

Each hematopoietic lineage appears to be specified by a unique combination of transcription factors that includes members of both groups[167-171]. The Notch signaling pathway was found to be required for the development of T-cells in the thymus[170].

The interaction of an antigen with TCR induces T-cell proliferation and differentiation and results in the conversion of T-cells from an immunologically inert state to an active state. All known signaling pathways emanating from the TCR are initiated by the Scr and Syk families of nonreceptor protein tyrosine kinase (PTKs).

The contact between presented foreign antigens and the antigen receptor of T-cells initiates signaling pathways that activate hundreds of genes responsible for coordinating the immune response.

Signalling through the T-cell receptor is required for the inititation of cellular immune response. The ligand for the TCR consists of a polymorphic class I or II major histocompitibility (MHC) protein complexes and antigenes (Baker et al., 1999). Formation of this complex play an important role for thymocyte development. The above fundamental data clearly demonstrate the importance of studying of molecular mechanisms of PR-1 action.

5.3. PROLINE-RICH-PEPTIDE IS A REGULATOR OF MYELOPOIESIS

It is known that among blood system pathologies a great place belongs to the leukopenic states with different genesis[172,173]. Leukopenia is a one of the basic indicators for formation of blood pathologies and the first hematological sign for many immunodeficient states[174]. Leukopenia can induce complications for a number of diseases, such as rheumatoidal arthritis, lupus erythematosis, hemoblastosis, viral and bacterial diseases, etc.[173,175]. However, the increased susceptibility to infections serves the basic manifestation of leukopenia[176-178].

At the present time a number of well-characterized biomolecules are known to stimulate myelopoiesis[179,180]. Nevertheless, the search for remedies capable of regulating myelopoiesis and preventing its consequences remains a challenge.

Earlier we discovered antibacterial activity of a new hypothalamic polypeptide that was termed PRP-1[116-119]. It was interesting to investigate its antibacterial action on the background of leukopenia.

The aim of the present study was to investigate the ability of PRP-1 to prevent the development of infectious processes caused by *P. aeruginosa* in mice with leukopenia.

Mice. Female pathogen-free BALB/c mice, 8 weeks of age and 14 g of body weight, received from the Laboratory of Animals "Stolbovaya" RAMS (Moscow, Russia) were used in the study. The mice were injected i/p with 4 mg/mouse of cyclophosphamide (CPA) for inducing leukopenia. 24 h later the mice were s/c injected with 10^{-6} g/mouse of PRP-1 or with an equal amount of saline alone (as a control).

Bacteria. P. aeruginosa, serotype 8, was received from the collection of I. I. Metchnikoff Research Institute for Vaccines and Sera (Moscow, Russia). The strain of bacteria was highly virulent and possessed typical cultural, morphological, and biochemical properties. The bacteria were cultured in tryptic soy broth (Difco Lab., Detroit, MI); overnight cultures were diluted serially with phosphate-buffered saline, and 500-μl samples of each dilution were injected into mice. Portions of suitable dilutions were also seeded onto agar plates for precise cell counting (CFU).

Determination of the lethal dose (LD) of P. aeruginosa for mice. Groups of 10 mice were i.p. injected with 10 different doses of *P. aeruginosa*, ranging from 1.25×10^3 to 2.5×10^6 CFU. The surviving mice were counted daily for 7 days, and the 50% and 100% LDs (LD_{50} and LD_{100}) were calculated by using Probit analysis[120].

Effect of PRP-1 on bacterial growth in leukopenic mice infected with P. aeruginosa. Four days after of CPA injection, mice treated s.c. with PRP-1 or with vehicle alone were i.p. injected with sublethal dose of the bacteria. At 6, 12, 24, 36, 48, 60, and 72 h after bacterial challenge, the mice were killed by cervical dislocation, and their spleens, livers, washes of peritoneal cavities, and peripheral blood samples were obtained aseptically. Each 100 μl of fluid from peritoneal cavities, peripheral blood, and spleen and liver homogenates were spread over nutrient agar. Colonies were counted after 18-h incubation at 37°C.

Effect of PRP-1 on survival rate of leukopenic mice infected with P. aeruginosa. Five days after the start of the experiments CPA-treated control mice and CPA and PRP-1-treated mice were i.p. injected with *P. aeruginosa* at different lethal doses. The surviving mice were counted daily for 7 days after the bacterial challenges.

Counting of cells. Peripheral blood cell counts and differential counts of leukocytes were performed using Giemsa-stained smears on the days 0, 1, 3, 5, 7, 9, 11, and 13.

Statistical analysis. Data were compared using the analysis of variance and by the Student's *t*-test.

After CPA administration, profound leukopenia was induced in both control and PRP-1-treated mice, and a nadir was reached on day 5 from the start of the experiments. However, the number of leukocytes in PRP-1-treated mice rapidly recovered after day 5 and was 2.8 - 7.2 times higher ($p < 0.001$) starting from day 7 than that in control mice, in which leukopenia lasted for a longer period (Fig. 37).

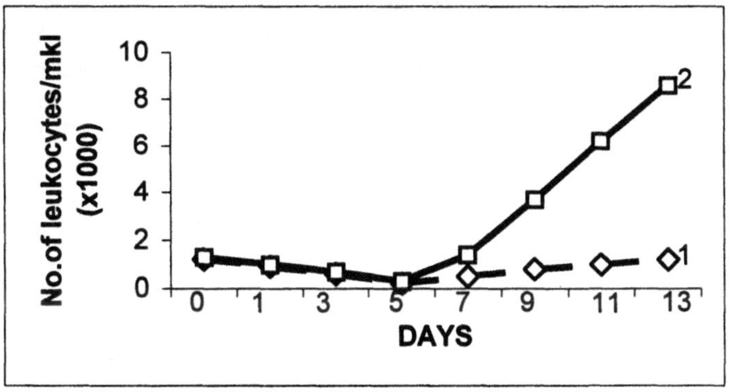

Figure 37. Effect of PRP-1 on peripheral blood leukocytes of CPA-treated mice.1 – control mice, 2 – PRP-1-treated mice.

Differential calculation of leukocytes showed that, starting from day 7, the leukocyte population in PRP-1-treated mice consisted mainly of neutrophils with segmented nuclei and monocytes. At the same time, the incremental ratios of neutrophils were higher than those of monocytes (Table 16).

Table 16. Effect of PRP-1 on the number of peripheral blood leukocytes in CPA-treated mice

Day	Treatment	Number of cells/μl				
		Neutrophils		Eosinophils	Monocytes	Lymphocytes
		Segmented	Banded			
0	none	210 ± 22	41 ± 7	12 ± 0.7	61 ± 4	1063 ± 51
1	Control	4 ± 0.3	1 ± 0.1	1 ± 0.1	3 ± 0.1	596 ± 25
	PRP	10 ± 0.7	2 ± 0.1	2 ± 0.1	6 ± 0.3	521 ± 22
3	Control	1 ± 0.1	1 ± 0.1	0 ± 0	5 ± 0.3	186 ± 8
	PRP	1 ± 0.1	0 ± 0	0 ± 0	8 ± 0.4	347 ± 14
5	Control	2 ± 0.3	2 ± 0.3	0 ± 0	1 ± 0.1	192 ± 8
	PRP	1 ± 0.1	0 ± 0	0 ± 0	2 ± 0.1	116 ± 6
7	Control	11 ± 2	3 ± 0.3	0 ± 0	9 ± 0.6	411 ± 18
	PRP	1013 ± 83	9 ± 0.8	15 ± 0.8	94 ± 6	436 ± 18
9	Control	57 ± 4	11 ± 0.9	3 ± 0.2	21 ± 1.8	684 ± 24
	PRP	5108 ± 112	111 ± 12	83 ± 6	613 ± 22	523 ± 21
11	Control	61 ± 4	16 ± 1.1	5 ± 0.3	26 ± 1.8	861 ± 36
	PRP	6421 ± 168	131 ± 12	94 ± 8	681 ± 28	783 ± 31
13	Control	73 ± 7	22 ± 1.3	6 ± 0.4	31 ± 2.2	936 ± 44
	PRP	7621 ± 213	133 ± 11	96 ± 8	744 ± 31	917 ± 42

Effect of PRP-1 on peripheral blood leukocytes in CPA-treated mice injected with P. aeruginosa. While the number of leukocytes in peripheral blood was gradually recovering starting from 12 h after infection both in PRP-1-treated mice and in control, their corresponding levels in PRP-1-treated mice were 1.3 ($p < 0.01$) – 3.9 ($p < 0.001$) times higher than those in controls (Fig. 38).

Differential cell counts revealed that the cells in PRP-1-treated mice consisted mainly of neutrophils with segmented nuclei at all sampling times, whereas in control mice the leukocytes increased in number consisted of both segmented and band-form nuclear neutrophils. The levels of monocytes of peripheral blood in PRP-1-treated mice were also higher than those in controls, at every experimental point (Table 17).

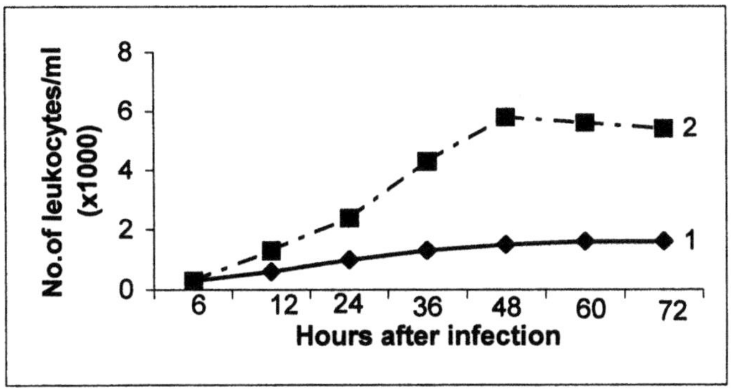

Figure 38. Changes in peripheral blood leukocytes in CPA-treated mice infected with *P. aeruginosa*. 1 – control mice, 2 – PRP-1-treated mice.

Table 17. Effect of PRP-1 on the number of peripheral blood leukocyte in CPA-treated mice infected with *P. aeruginosa*

Time post infection (h)	Treatment	Number of cells/μl				
		Neutrophils		Eosinophils	Monocytes	Lymphocytes
		Segmented	Banded			
6	Control	1 ± 0.1	0 ± 0	0 ± 0	0 ± 0	151 ± 7
	PRP	22 ± 0.8	7 ± 0.4	0 ± 0	16 ± 0.6	168 ± 8
12	Control	36 ± 3	13 ± 0.5	0 ± 0	5 6 ± 4	359 ± 15
	PRP	486 ± 21	54 ± 3	0 ± 0	176 + 8	267 ± 11
24	Control	42 ± 2	33 ± 3	0 ± 0	81 ± 4	633 ± 26
	PRP	662 ± 26	111 ± 5	0 ± 0	384 ± 15	388 ± 16
36	Control	114 ± 5	88 ± 4	0 ± 0	112 ± 5	711 ± 29
	PRP	2668 + 111	126 ± 6	0 ± 0	448 ± 18	366 ± 15
48	Control	264 ± 12	161 ± 8	0 ± 0	146 ± 7	862 ± 33
	PRP	3026 + 131	243 ± 6	0 ± 0	592 ± 23	363 ± 14
60	Control	445 ± 18	283 ± 11	0 ± 0	76 ± 3	532 ± 21
	PRP	2933 ± 126	121 ± 6	28 ± 1	404 ± 16	412 ± 16
72	Control	613 + 23	344 ± 14	0 ± 0	48 ± 2	411 ± 17
	PRP	2886 ± 124	94 ± 4	24 ± 1	278 ± 11	444 ± 19

Effect of PRP-1 on leukocytes of peritoneal cavities in CPA-treated mice infected with P. aeruginosa. Six hours after bacterial challenge the number of leukocytes in peritoneal cavity fluid samples from PRP-1-treated mice was significantly higher than in the samples from control mice. However, the number of leukocytes in control mice samples significantly exceeded that in PRP-1-treated mice samples at 24 h and thereafter at all time points (Fig. 39).

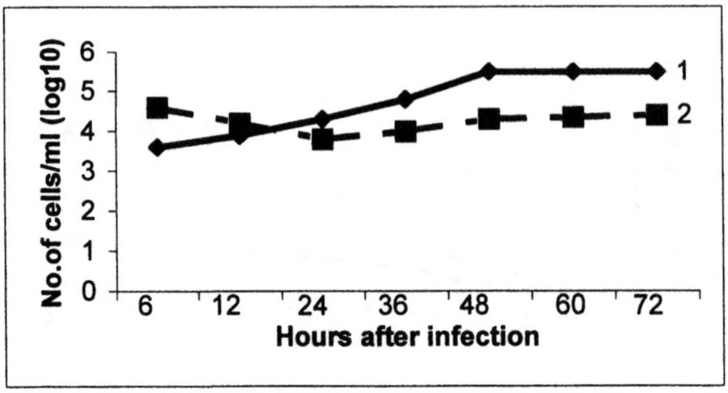

Figure 39. Changes in peritoneal fluid cells in CPA-treated mice infected with *P. aeruginosa*. 1 – control mice, 2 – PRP-1-treated mice.

Differential cell counts revealed that peritoneal fluid cells in PRP-1-treated mice consisted mainly of neutrophils with segmented nuclei and macrophages at all sampling times. On the other hand, the cells in 6-h peritoneal cavity fluid samples from control mice were mainly lymphocytes and band–form neutrophils which number gradually increased with time (Table 18).

Table 18. Effect of PRP-1 on leukocyte number in peritoneal cavities of CPA-treated mice infected with *P. aeruginosa*

Time post inoculation (h)	Treatment	Number of cells x 10^3/ml				
		Total	Neutrophils		Macrophage	Lymphocyte
			Segmented	Banded		
6	Control	68 ± 4	3.7 ± 0.3	0.3 ± 0.1	6.6 ± 0.3	52 ± 2.1
	PRP	352 ± 25	102 ± 4.3	16 ± 0.4	86 ± 4.2	125 ± 4.8
12	Control	134 ± 8	14 ± 0.5	7.8 ± 0.2	16 ± 0.4	164 ± 6.6
	PRP	72 ± 5	11 ± 0.4	6.4 ± 0.2	3.8 ± 0.2	82 ± 3.2
24	Control	286 ± 16	21 ± 0.8	13 ± 0.5	25 ± 1.1	234 ± 9.8
	PRP	76 ± 5	11 ± 0.5	1.6 ± 0.1	4.1 ± 0.3	61± 2.4
36	Control	1880 ± 86	765 ± 31.3	144 ± 5.8	189 ± 7.4	834 ± 29
	PRP	486 ± 21	214 ± 7.4	17 ± 0.4	83 ± 8.3	114 ± 4.8
48	Control	3060 ± 130	1162 ± 58	257 ± 11.1	313 ± 12.3	1029 ± 52
	PRP	654 ± 32	349 ± 14	25 ± 0.8	115 ± 3.4	170 ± 5.6
60	Control	2740 ± 121	1166 ± 56	365 ±14	324 ± 12.6	666 ± 24
	PRP	612 ± 31	387 ± 16.4	21 ± 0.8	85 ± 8.8	64 ± 2.6
72	Control	2550 ± 116	1189 ± 59	417 ± 17.6	349 ± 14.6	486 ± 18
	PRP	563 ± 26	457 ± 21.6	17 ± 0.4	77 ± 8.4	23 ± 0.8

Effect of PRP-1 on the persistence of bacteria in internal organs of CPA-treated mice infected with P. aeruginosa. The mice were injected by sublethal doses of bacteria. The number of viable bacteria in the body of PRP-1-treated mice was reduced, and the elimination of bacteria from the body was significantly enhanced (Fig. 40). The largest significant differences between PRP-1-treated mice and control mice were observed at 24 h after infection.

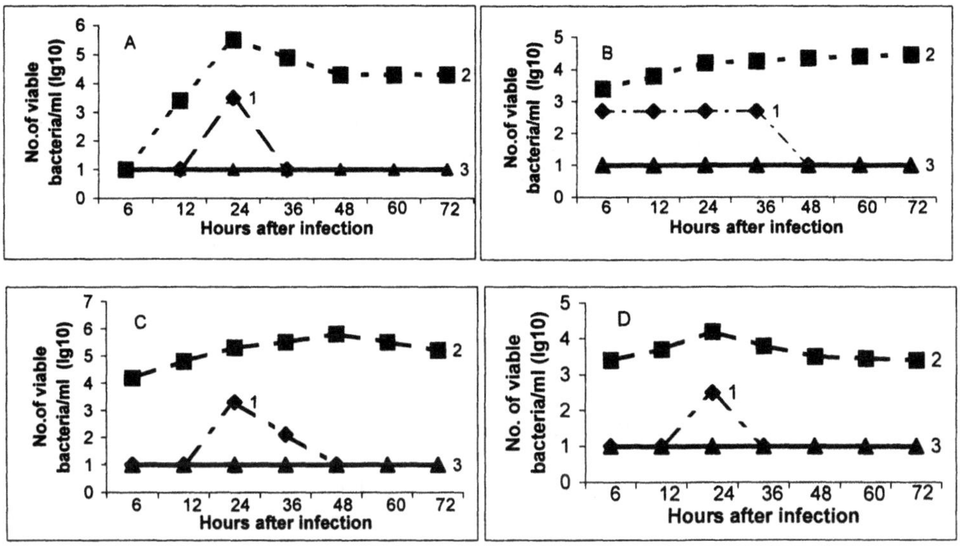

Figure 40. Effects of PRP-1 on the number of viable bacteria in peritoneal cavities (A), spleens (B), livers (C), and peripheral blood (D) of CPA-treated mice infected with *P. aeruginosa*. 1, PRP-1-treated mice; 2, control mice; 3, undetectable area.

Effect of PRP-1 on survival rate of CPA-treated mice infected with P. aeruginosa. CPA-treated mice were injected with different lethal doses of bacteria, and the number of surviving mice was determined. PRP-1 significantly enhanced survival of mice and had a pronounced protective effect during the period of development of the infection caused by bacterial agents. In particular, when mice were injected with 1.2 LD_{50} of bacteria, survival in control mice was 20%, but in PRP-1-treated mice it was 70%. When mice were injected with 1.2 LD_{100} of bacteria, survival in PRP-1-treated mice was 30%, whereas no survivors were observed among control mice given vehicle alone (Fig. 41).

In the present study we modeled leukopenia and studied quantitative and differential displacement of blood leukocytes; next, we investigated the influence of PRP-1 on these processes, as well as its ability to prevent the development of infection induced by highly virulent representative of gram-negative bacteria, *P. aeruginosa*, on the background of leukopenia in mice.

In our experiments, CPA injected i/p into mice at a dose of 4 mg/mouse produced classical leukopenia. 24 h after CPA injection the mice were treated s/c with a single injection of PRP-1, 10^{-6} g/mouse. In our previous study the ability of PRP-1 to enhance the accumulation of phagocytes in the peritoneal cavities of mice after i/p administration was shown[117]. We injected PRP-1 into mice by the s.c. route in order to avoid this effect.

On the 5th day of observation the lowest levels of peripheral blood leukocytes were registered in CPA-PRP-1-treated mice and CPA-treated control mice. There was no significant quantitative difference between PRP-1-treated and control mice. However, the level of neutrophils in the blood of PRP-1-treated mice started growing from day 7 and increased significantly during subsequent period of observations (Fig. 36, Table 16). It is well known that *P. aeruginosa* is one of the main pathogens responsible for purulent septic infections[181].

Also, it is known that protection against *P. aeruginosa* is provided mainly by neutrophils[181,182]. We considered day 5 after injection of CPA as an appropriate time for administration of *P. aeruginosa* to mice because that was the time of lowest content of neutrophils in the peripheral blood. It was shown that PRP-1 prevents the development of the infectious process by means of recovery of neutrophil content in peripheral blood and peritoneal cavities of mice with leukopenia. It is interesting that the number of both neutrophils and total leukocytes in peripheral blood samples from PRP-1-treated mice, even 24 h after infection, was significantly higher than in controls and showed a similarity to that in mice without *P. aeruginosa* infection (Fig. 39, Table 17).

The number of neutrophils in the fluid from the peritoneal cavities of PRP-1-treated mice was significantly higher than in controls 6 h after i.p. inoculation of *P. aeruginosa*. Thereafter, however, the numbers of neutrophils in peritoneal cavity fluid samples from PRP-1-treated mice were consistently lower than those in fluid samples from control mice (Fig. 40) and correlated to a rapid decrease in the numbers of bacteria in the former (Fig. 41).

It should be taken into account that the number of bacteria in internal organs of PRP-1-treated mice was lower and that PRP-1 reduced the duration of bacterial persistence in animal body and enhanced their elimination from the body (Fig. 40). The phenomenon noted above may reflect the healing of infection after bacterial killing by neutrophils and may be due to the generation of superoxide by mature neutrophils by oxygen-dependent mechanisms. Remarkably, PRP-1 caused an increase not only in the number of neutrophils, but also in the fraction of more mature neutrophils bearing segmented nuclei (Tables 16 and 17).

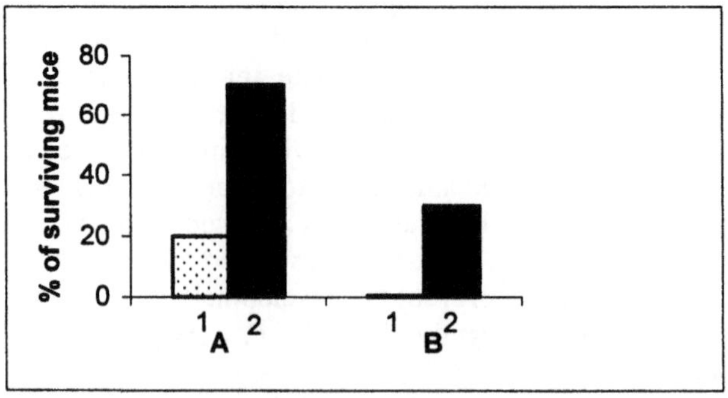

Figure 41. Effect of PRP-1 on the survival rate of CPA-treated mice infected with *P. aeruginosa*. A – 1.2 LD_{50}, B – 1.2 LD_{100}. 1 – control mice, 2 – PRP-1-treated mice.

The present study revealed that PRP-1 possesses a strong protective effect against infection caused by *P. aeruginosa* in mice with leukopenia. Prophylactic treatment with PRP-1 protected mice from usually lethal infections, enhancing their survival (Fig. 42).

Our present observations suggest that PRP-1 contributed to the protection against *P. aeruginosa* infection in mice with leukopenia by enhancing the production and functional activity of neutrophils, especially those with segmented nuclei.

It is well known that neutrophils are formed by myeloid graft of blood formation from blood stem cells along the myelocytic pathway termed myelopoiesis[175,183]. Our results indicate the ability of PRP-1 to bring about rapid recovery of myelopoiesis. PRP-1 may be a potentially useful adjunct in clinical medicine for the protection of immunocompromised hosts against serious *P. aeruginosa* infection. Moreover, PRP-1 may be of help in acute and chronic neutropenia, decreased myeloid progenitor frequencies in bone marrow and spleen, and increased susceptibility to bacterial infection.

Molecular mechanisms of PRP-1 action in myelogenesis can be explained by stimulation of expression of granulocyte colony-stimulating factor (G-CSF), which is a potent and relatively specific stimulator of neutrophil production *in vivo*. A number of observations suggest that G-CSF play an important regulatory role in granulocytopoiesis. G-CSF uses a receptor of the cytokine receptor superfamily and induces tyrosine phosphorylation and activation of the Janus protein kinase (Jak) family members. G-CSF induces tyrosine phosphorylation and activation of Jak1, Jak2, and Tyk2 kinases, and subsequently activates a family of transcription factors including STAT3 (signal transducer and activator of transcriptor) as well as STAT1 and STAT5[184,185]. Our suggestion is based on our pervious data regarding the effect of PRP-1 on the biosynthesis of GFAP (glial fibrillary acidic protein) in astrocytes by stimulation of the TrkB (family of tyrosine kinases) receptor[186].

It is known that transcription factors play the major role in the differentiation in a number of cell types, including the various hematopoietic lineages. The roles of myc, myb, MZF-1, and PU.1 transcription factors in myeloid development have been studied. Recent studies have focused on PU.1 as a major regulator of myeloid development. PU.1 knockout mice failed to produce certain myeloid lineages, suggesting that Spi-B might be expressed quite differently from PU.1. Transcription factor AML 1 is normally expressed in hematopoietic tissues and during myeloid differentiation is also expressed in nervous tissue, skeletal muscle, and reproductive tissues. Interaction of PU.1 with AML 1 produced by nervous tissue is important for myeloid gene expression[186-190]. One of the molecules influencing gene transcription may be peptide PRP-1 produced by NPV and NSO.

PU.1 represents a unique transcriptional regulator within the hematopoietic system as it is essential for the generation of both myeloid (macrophages, neutrophils, mast cells) and lymphoid (B- and T-lymphocytes) lineages. PU.1 functions in a cell-autonomous manner controlling the proliferation and differentiation of lymphoid-myeloid progenitors. PU.1 controls the proliferation of these progenitors by regulating the expression of hematopoietic cell-specific cytokine signal-transducing molecules such as interleukin-7 receptor (IL-7R), macrophage colony-stimulating factor receptor (M-CSFR), granulocyte CSF receptor (G-CSFR), and granulocyte-macrophage-CSF receptor (G-MCSFR). At least three adhesion sites of fibronectin, (FN-C)HI, (FN-C)HII, and CS I, are involved in the interaction with colony-forming cells (CFC). These peptide fragments of fibronectin contain 5-7 proline residues. The progenitor cells recognize fibronectin[189].

The cytokines are thought to produce their effect on the genes through transcription of STAT. Four STATs have been described up to date: STAT1, STAT2, STAT5, and STAT6. STAT transcription factors can be activated by phosphorylation of certain tyrosine residues[184,191]. As shown, PRP-1 possesses IL-8-like properties such as activation of neutrophils. Myeloid and erythroid precursors, as well as their progeny, reach and exceed normal levels in Ikaros null mice. In contrast to the reduction of hematopoietic stem cell (HSC) activity and severe blocks of differentiation toward most lymphoid lineages, a dramatic increase in myelopoiesis takes place.

Ikaros also has a pivotal role at subsequent stages of the hematopoietic pathway in determining the lymphoid versus of erythro-myeloid choice of differentiation. Thus, Ikaros can be considered as a positive regulator of HSC and lymphoid precursor production and as a negative regulator of myeloid differentiation[190,192]. Our data demonstrated that PRP-1 is a stimulator of myeloid as well as lymphoid precursors. Thus, PRP-1 in this respect is a unique regulator of lympho-myeloid differentiation[33,34,36].

The effect of PRP-1 is similar to that of PU.1; the latter plays a critical role in the generation of both lymphoid and myeloid lineages. It is known that PU.1 transcription factor regulates the survival of these populations by selecting a growth factor receptor to be expressed. For the maintenance of myeloid cell growth, apart from hematopoietic growth factors G-CSF, M-CSF, GM-CSF, Epo, Tpo, IL-3, and IL-5[173,176,180,183,193], an important role is played by IL-6, IL-11, LIF (leukemia inhibitory factor), oncostatin M, ciliary-neurotrophic factor, and cardiotrophin-1 as well as hematopoietin receptor[194-196], IL-1, IL-9, and stem cell factor (SCF/c-kit ligand)[183,193].

Peptide regulators of myelopoiesis were recently isolated from the bone marrow[197]. Our findings indicate the importance of studying the molecular mechanisms of PRP-1 effect on myelopoiesis as well as lymphopoiesis. Because Ets factors such as PU.1 and c-myc both play an important role in myeloid development, it is likely that these factors interact with AML 1. Phosphorylation of AML 1 by signal-regulated protein kinase ERK 1 at two serine residues within the proline-serine-threonine-rich region in its carboxyl terminal domain is important for myeloid-specific gene expression. Several lines of evidence indicate a key role for the retinoic acid receptor (RAR) in myeloid differentiation: vitamin A deficiency is associated with defective hematopoiesis, and retinoids stimulate granulopoiesis. The homeobox genes in myeloid progenitor cells may also be important for their maturation. There are other factors involved in myeloid development such as c-myb (isolated from retroviral oncogen), c-myc, NF-KB, p53 protein, and c-Jun proto-oncogene.

Taking into account that T-lymphocyte differentiation is revealed on day 13 of mouse embryonic development, before appearance of $CD4^+$ and $CD8^+$, one may suggest that PRP-1 also has an influence on bone marow function. Our data about localization of substantial quantity of PRP-1 in bone marrow ganules support this suggestion. So, we have to take into consideration that PRP-1 enhanced the number of cells 1.23 - 3.1 times ($p < 0.001$) in vivo and 1.34 - 3.5 times ($p < 0.001$) in vitro on the 13-19th days of gestation. Our investigations on PRP-1 influence on pluripotent stem cells of the bone marrow are in progress.

Thus, PRP-1 can be considered as a novel factor of differentiation of T-cells, their formation from corresponding progenitors, and also a factor of myelopoiesis. The biological activity of PRP-1 was demonstrated both during the embryogenesis and postnatal animal development. In general, PRP-1 is one of important hematopoietic factors.

6

NEUROPROTECTIVE (ANTINEURODEGENERATIVE) PROPERTIES OF PRP-1

6.1. PROTECTIVE EFFECT OF PRP-1 AGAINST SNAKE VENOM-INDUCED NEURONAL INJURY

At the present time toxic proteins of snake venom are grouped into more than 10 categories in accordance with chemical structure and pharmacological properties. The following 4 big groups have been allocated by Karlsson[198]: curaremimetic postsynaptic neurotoxins, a multitude of membrane toxins, toxins with phospholipase activity, and crotamine.

In addition, snake venoms contain considerable amount of enzymes, which are used in biochemical studies and in medicine. The quantity of such enzymes has reached 26, most of them being hydrolases. Twelve out of them were found in all venoms studied, although their content differ significantly. It is generally accepted that snake venoms cause local capillary damage and tissue necrosis through the action of proteinases, phospholipases, etc.[199-201]; they cause diverse coagulant and anticoagulant effects by the action of various proteinases and phospholipase A[202]; they also induce acute hypotension and pain due to release of vasoactive peptides by kininogenase[201]. At the same time, it has been demonstrated that none of these enzymes is responsible for the acute toxicity of snake venoms[203]. Some components of snake venoms became popular and are used in general anesthesia, as well as synthetic tubocurarine analogues, atracurium[204,205] and vecuronium[206,207].

Snake venom toxins are of extreme importance also as inhibitors of angiotensin-converting enzyme (ACE), responsible for converting an inactive precursor into the locally active hormone angiotensin, which causes the elevation of blood pressure. Moreover, it was revealed that the inhibitor of ACE is capable of dilating blood vessels by prolongation of the action of bradykinin, another locally active hormone, by the mechanism of blocking and inactivating the enzyme[208]. As a consequence, pharmacological blocking of the enzyme can reduce blood pressure. Analogs of ACE inhibitors are captopril, enalapril, and lisinopril. Future development of snake toxin pharmacology is connected with drugs called dendrotoxins. They were first found among dendroaspis venom protein toxins, on the basis of which important analogs were synthesized that selectively block or activate the same neuronal potassium ion channels by which the neural excitability is controlled. Blockers of K^+ channels can increase the activity of damaged nerve cell in neurodegenerative disorders (AD), and activators of K^+ channels are able to reduce abnormal electrical activity of the brain and to be useful as anticonvulsant drugs for epilepsy[209].

The fact of active interaction of snake venoms with hormones, as well as their action on regenerative processes in the nervous system served as the starting point for our recent studies of PRP-1 effect on SC neurons subjected to the action of the Armenian adder (Vipera Raddei) venom, in normal and pathological (hemisection of SC) conditions. Recently an analogous study of the combined action of this venom and dexamethasone (DEX) has been reported[210]. However, an advantage of the present investigation is the combination of snake venom with the native hormone, PRP-1. Moreover, besides the study of the PRP-1 effect on the autonomic nervous system, an investigation of PRP-1 influence on the somatic nervous system, both its independent action and that in combination with the snake venom, in normal and pathological conditions, was performed[210].

Protective effect of PRP-1 against Vipera Raddei venom-induced neuronal injury. The experiments were performed on 27 adult male rats (Wistar, 200-300 g). The rats were restrained by ditiline and transferred to artificial respiration. SC was cut under anesthesia (Novocaine) by an ultrasonic knife on levels T_2-T_3. In the lumbo-sacral part of the spine a laminectomy was made after fixation of this region in a stereotaxic apparatus. A group of rats (over 3-4 weeks) were subjected to preliminary hemisection of SC on levels L_2-L_3. Extracellular background activity (BA) of SC single neurons was investigated under the influence of PRP-1 (from 35 to 100 μg) and Vipera Raddei (VR) venom which was injected i/m at the doses of 0.5 and 1.3 mg/kg, the toxic dose being 0.7 mg/kg. Lyophilized crystal application was also used. The recording of BA was carried out by glass microelectrodes (tip diameter 2-6 μm) filled with 2 M NaCl solution and stereotaxically oriented in dorsal regions of SC gray matter (III - VI laminae by Rexed). Registration of field potentials (FP) in the dorso-ventral direction of SC was also conducted. The registration was made using a computer with the help of a special program enabling on-line spike discrimination and selection. The subsequent output was averaged, and the pulse histogram permitted the choice of a necessary mode of evoked single interneuron (IN) or motoneuron (MN) activity.

Data analysis was performed with use of a known algorithm[210-212]. The glided frequency by 10 interimpulse intervals with a spacing of 5 intervals, serial correlogram to 50 orders, and the histogram of interimpulse intervals (HII) before and after 1 – 5-sec stimulation were calculated. On the basis of the PETH (Peri-Event Time Histogram) interspike, intervals of the poststimulus neuronal tonic activity were calculated by means of consecutive truncation of examined spike flow by exception of band values with spike amount relapsing from interval M ± 2 SD, with the repeating of this procedure until both M and SD values were stabilized (distinction less than 2 %). The final sample was accepted as neuronal tonic activity. By criterion a χ-square its conformity normal, Poison or uniform distribution was estimated. With dependence of distribution the significance of quantity spike of poststimulus histograms where they fell outside the limits of M ± 2SD was calculated. The significance of distinction of BA and tonic activity was determined with the help of two-tailed Student's *t*-test. A similar problem was solved by means of the cumulative curves constructed for BA and tonic histograms. By the least squares method the inclination of the straight lines approximating the given curves was defined. On a slope of the straight line approximating the given site of a cumulative curve, the probability of the significance of a divergence paid off with tonic activity was calculated.

The effects of PRP-1 on spinal rats and on those with preliminary hemisection of the SC (over 3 weeks) were studied independently and in combination with VR venom treatment, taking as an example the FP and the impulse flows of 47 single neurons of SC (laminae III-VI by Rexed) of BA[210].

Fig. 42 shows a significant depression of BA under the influence of VR venom (*A*), a pronounced activation of FP under the action of PRP-1 (*B*), and the absence of venom effect on the background of PRP-1 (after its action for 60 min, *C*). The same is demonstrated as a diagram in Fig. 42*E* for the above mentioned cases.

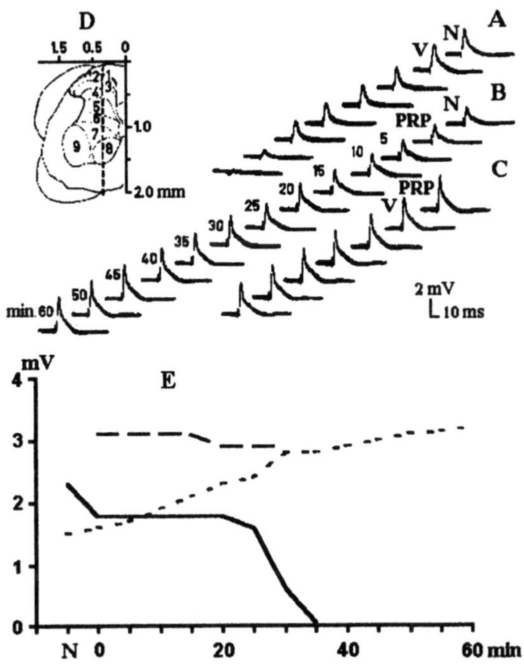

Figure 42. Separate and combined influence of the PRP-1 and VR venom on FP of the SC IN (depth 500 μm). A - C - influence of venom - V (A), PRP-1 (B) in norm - N and V on the background of PRP-1 preliminary action (C). C - continuation of B. D - schematic delineation of SC on level T5 - T6; numbers indicate the laminae of SC gray matter by Rexed; dotted line - direction of microelectrode's insertion; vertical and horizontal scales - width and depth of the SC in mm. E - graphic representation of effects in A - C: continuous line (A), intermittent line (B), intermittent with points lines (C). Ordinate and abscissa - amplitude in mV and time in min after giving drugs, respectively; N indicates initial values of FP before influence of V and H.

Fig. 43 shows the action of different doses of PRP-1 on two background-active SC neurons (*A, B*). Depending on the starting level of BA (*A1, B1*) and the hormone concentration (*A* – 70 μg, *B* - 100 μg), different degrees of neuron activation under the influence of PRP-1 are revealed, more pronounced at smaller PRP-1 concentration and low starting frequency (*A*).

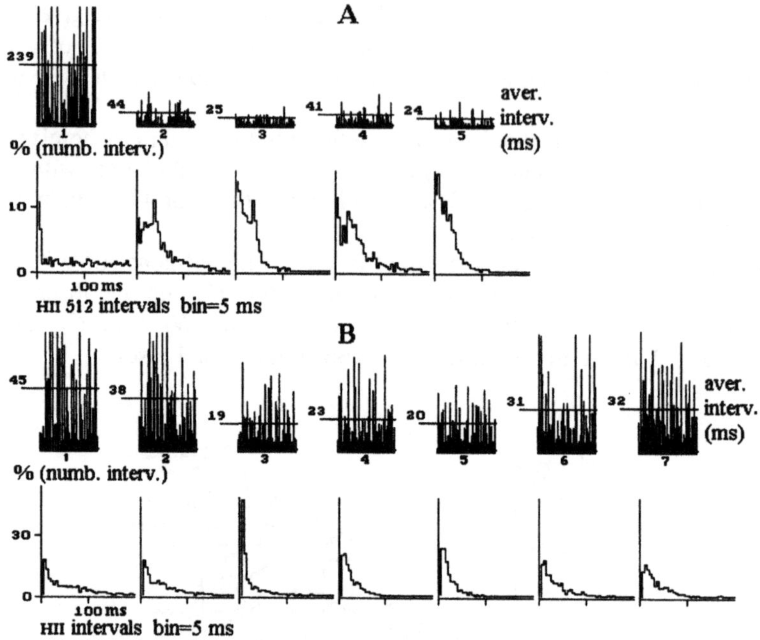

Figure 43. Effects of PRP-1 different doses on BA of SC neurons. The alterations of time parameters of neuronal flows of two neurons (depth 600, 700 μm - A, B, respectively): in form a select arbitrary piece of short period of BA impulsion with indication of average means of interimpulse intervals (numbers from the left) on vertical (upper row) and histograms (bottom row) of the quantity of interimpulse intervals in percent from common number of those (HII) under systemic (i/m) injection of 70 γ (A), 100 γ (B) for electrical activity in normal state (A1, B1) with subsequent registration of PRP-1 effects at 5, 15, 25, 35 min (A2 - A5), 5, 10, 15, 20, 25, 30 min (B2 - B7).

Fig. 44 presents the graph of activating action of PRP-1 on BA of three spinal neurons within 30 min. Significant enhancement of frequency impulsion to 20 -30 min takes place.

Figs. 45 and 46 demonstrate the analysis of SC neuron single impulse flows under conditions of N. ischiadicus (I) stimulation in normal state (Fig.45), after 30 min of venom action (Fig. 46 *A - C*), and in the period from 2 to 30 min (summarized pre- and poststimulus histogram of interimpulse intervals) (Fig. 46*D*).

In Fig. 45 after the stimulation an increase of cumulative curve initial angle (*A2*), which then assumes the form of comparatively stationary activity, somewhat lowered as compared with BA (*A1*).

In both cases (Fig. 45*B*, by graphs of glided frequency and serial correlogram, and Fig. 46) through a short period after stimulation (5 sec), the restoration of the stationarity corresponding to pre-stimulus flow of spikes is shown.

In Fig. 46 a considerable increase in frequency impulsion of paroxysmal type is seen despite the fact that the pre-stimulus activity has been significantly inhibited as compared to normal conditions (*A*). This is also obvious from Fig. 47*C*, where an analogous effect is demonstrated on examples of summation of those from two neurons.

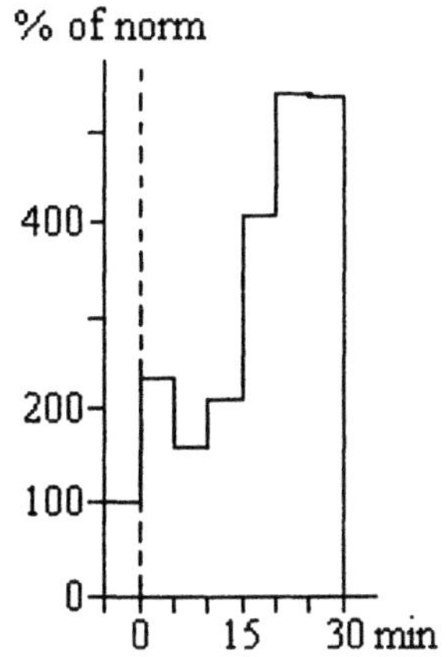

Figure 44. The graph of PRP-1 effects on average frequency of neuronal BA at 5 to 30 min exemplified by three SC neurons (III-VI laminae by Rexed). Ordinate, average frequency of BA relatively to normal state (100%); absciss, time of recording after administration of PRP-1 (vertical line).

Besides, Fig. 47 summarizes histograms of BA interimpulse intervals: before and after the stimulation of I nerve in normal conditions (*A*); two distinctive types of changes of the interimpulse intervals influenced by different venom doses (small - *B, D* - and large - *C*), providing more pronounced effect of venom under small doses also. Moreover, as is indicated on *B* and *C*, the venom evoked initial excitation and subsequent inhibition (*B*) on the one hand and pronounced facilitation of activity in later poststimulus periods (*C*) on the other, as opposed to PRP-1 effect (*D*). The last is characterized by the absence of pronounced changing of excitability of neurons, revealed in poststimulus period, by summation of data of all the trials from 5 to 60 min (*D*).

Of some interest is a pronounced effect of a small dose (35 μg) of PRP-1 even under conditions of paroxysmal bursting of neuronal activity upon preliminary application of a maximal dose of the venom (Fig. 48*A*). In this case a considerable (8-fold) BA decrease and gradual disappearance of bursts is observed. Thus, again the adaptive effect of PRP-1 takes place, even in its minimal dose, in comparison with the preceding case (Fig. 43). In this case high efficiency of PRP-1 small doses is observed. The absence of effect of systemic venom injections is revealed after preliminary injection of PRP-1 (Fig. 48*B*). The application of the venom revealed only 2-fold decrease in BA frequencies without pronounced initial excitatory effect, the result confirming protective properties of PRP-1.

Of special interest is the effect of PRP-1 on injured nerve cells under conditions of preliminary SC hemisection (Fig. 48*C*). As a rule, in this case, on injury side, under the section, multiple decrease in the level of BA frequency takes place (Fig. 48*C*, 2). After 10-min action of PRP-1 the rehabilitation of activity to the normal level and even to a level 2 times higher takes place (compare 1 and 5 in Fig. 48 *C*), which makes evident the great perspective of PRP-1 use for clinical practice.

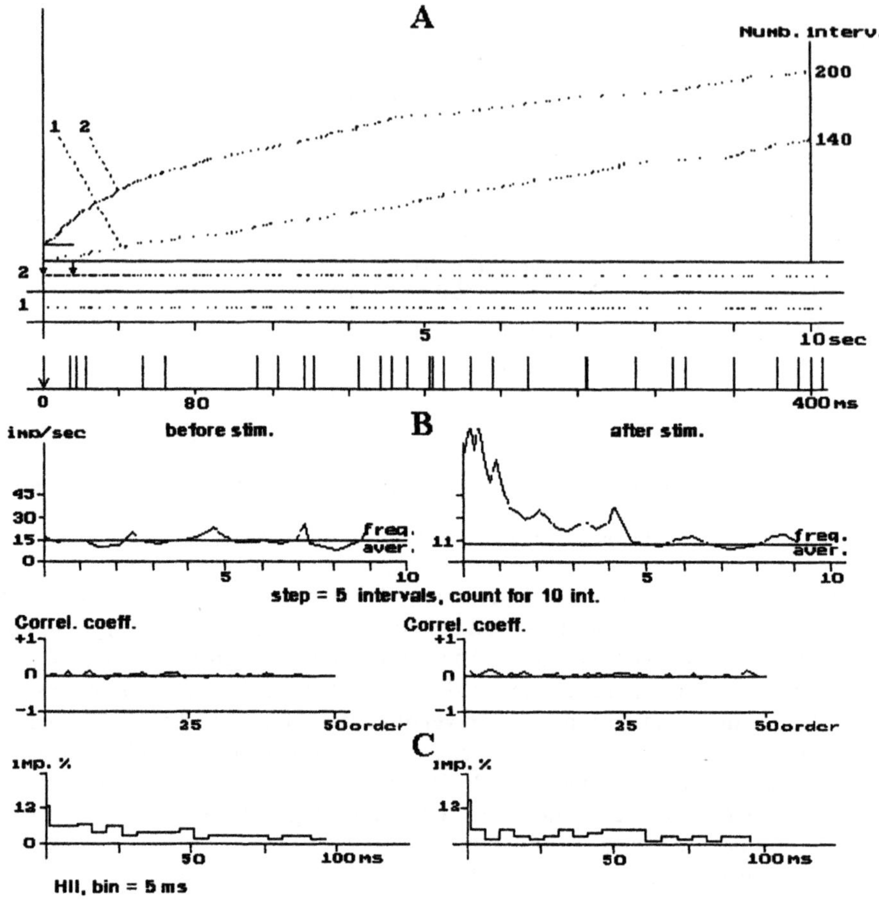

Figure 45. Pre- and post-stimulus delineations of a background active single SC neuron impulse flow in normal state (depth 700 μm). A - interimpulse intervals in real time before (1) and after (2) stimulation and their cumulative histograms; statistic parameters of impulse flows are given, below - unfolded picture of spike activity in selected intervals after stimulation. B - from above the graphs of glided frequencies impulse flow before and after stimulation (the levels of average impulsion frequencies are shown), below - the serial correlograms before and after stimulation. C - histograms of interimpulse intervals in percent of their common number before and after stimulation.

Fig. 49 demonstrates the stimulatory effect of PRP-1 on BA frequency in normal state and in the case of SC hemisection.

Next series of our experiments was performed to study the effects of PRP-1 and of a strong corticosteroid hormone DEX on SC neurons in normal conditions and in pathological state caused by VR venom and SC hemisection.

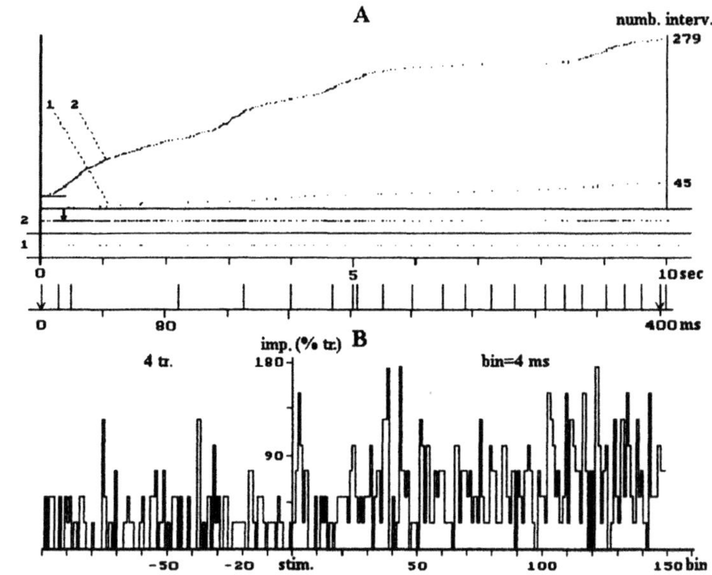

Figure 46. Characteristics of neuronal impulse flow subjected to the action of a small venom dose. A - the same neuron, as in Fig. 45, after 30 min of venom action. B - summarized histogram of neuronal activity before and after stimulation (from 2 to 30 min).

For estimation of neuroprotective properties of tested hormone we used models of acute and chronic neurodegeneration [204,205,210-214].

In experiments when SC hemisection was accompanied by daily injections of DEX during post-operational survival period (up to 3 weeks), the FP average amplitude was reduced by 40% at stimulation of I nerve on the injured side below the place of section, in comparison with normal state. DEX injection in these conditions induced an insignificant increase (by 8 %) in FP amplitude. On the control side of SC the reduction of FP amplitude was less expressed (on the average by 28 %) in comparison with normal state; DEX injection lead to a raise of FP amplitude to the level of normal conditions. On the injured side of SC, in contrast to DEX, PRP-1 exerts a powerful effect on FP amplitude in comparison with the initial level in normal state accepted as 100%. After PRP-1 injection the average FP amplitude rises, exceeding the initial value at 10 min by 6%, and at 60 min by 30%, while the PRP-1 effect on FP under normal conditions is more pronounced: at 60 min the initial average amplitude is exceeded by 115 %.

Thus, the hormone effect is less pronounced on the injured side of SC in comparison with non-injured side. Besides, the PRP-1 effect on the injured side is stronger in contrast to DEX effect (30% and 8%, respectively).

The above experiments show that PRP-1 efficiency is higher in comparison with DEX. DEX causes only two-fold increase in the frequency of neuron BA on the injured side, the level of average frequency approaching to normal state, but not reaching the level of the control side. At 10 min after PRP-1 injection its effect already differs by one order from that in control.

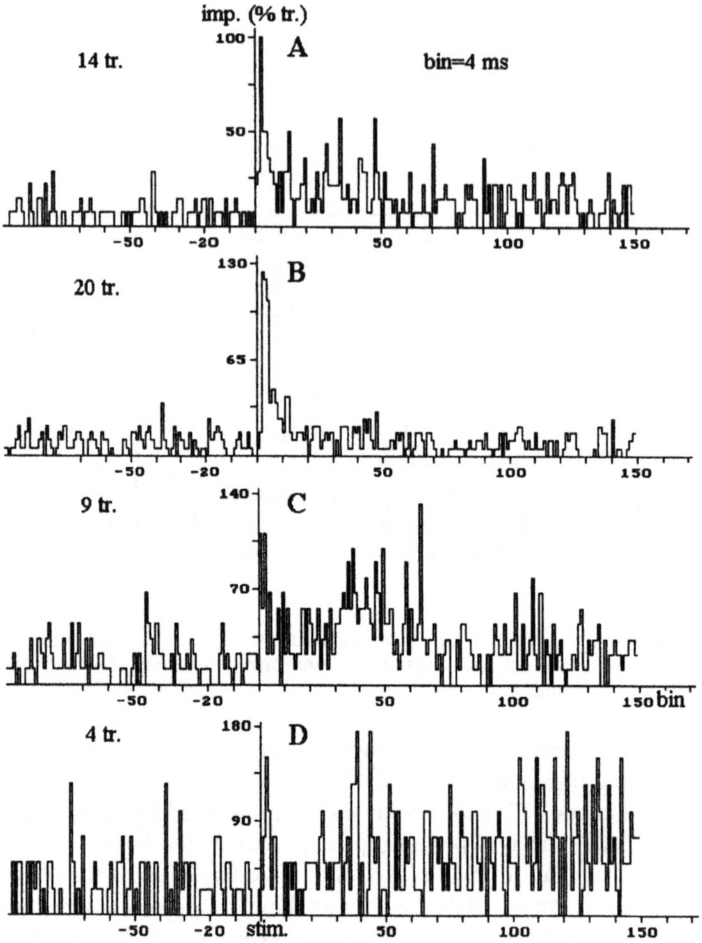

Figure 47. Summarized pre- and poststimulus histograms of inter-impulse intervals of SC single background active neuron's in normal state (A), under influence of low (B, D) and high (C) venom doses (laminae III - VI by Rexed).

So, on the injured side, below the site of section, a multiple decrease in BA frequency level takes place. Therefore, 10 min is enough for PRP-1-induced restoration of neuronal activity up to the normal level, the fact suggesting a great perspective of its use in clinical practice.

Protective effect of PRP-1 against cobra venom-induced neuronal injury. The goal of our next study was to investigate the effect of PRP-1 on cobra venom (CV)-damaged SC.

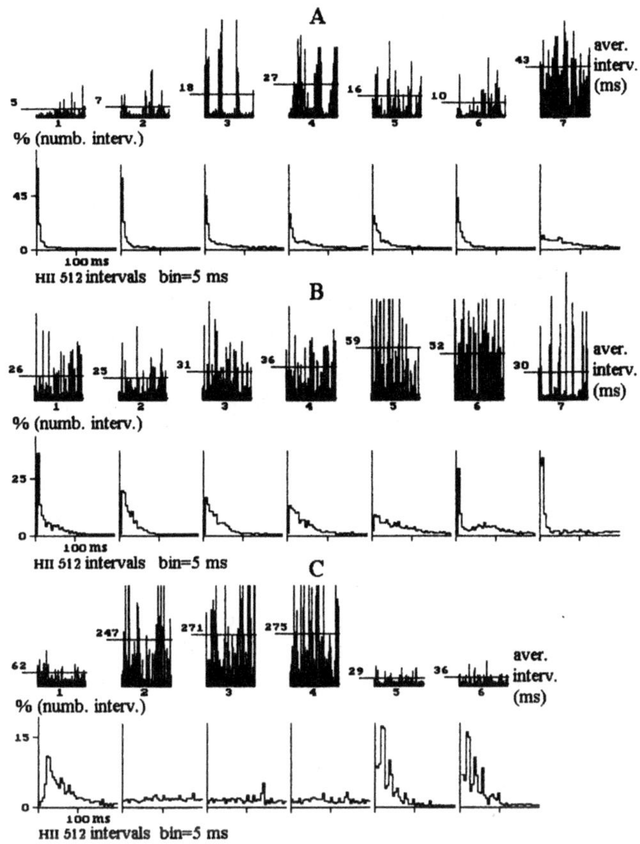

Figure 48. Influence of venom on BA of SC neurons after preliminary PRP-1 injection (depth 500 and 850 μm), and also influence of PRP-1 on BA in neuron injury resulting from SC hemisection (depth 1000 μm). A - C - BA of three neurons. The activity in normal state (A1 - C1); after hemisection (C2, C3); 5 and 10 min after preliminary venom application (A2, A3) followed by registration of PRP-1 effect in 2, 10, 15, and 20 min (A4-A7); 5, 10 and 15 min. after preliminary PRP-1 injection (B2 - B4), 5, 10 min after venom injection (B5, B6) and PRP-1s subsequent application (B7); 5, 10, 15 min after PRP-1 action on the activity of injured neurons below the section site (C4 - C6, respectively).

The experiments were carried out on 42 adult male rats (Wistar, 200 - 300 g) in a series of acute neurodegenerations (CV intoxication). The SC was cut with an ultrasound knife at T_2-T_3 level. After fixation of the lumbosacral region in stereotaxic apparatus the laminectomy was performed. The rats were immobilized by ditiline and artificially ventilated. The single background and evoked spike activity (after stimulation of flexor n. peroneus communis (P), extensor n. gastrocnemius (G) and mixed n. tibialis of hind limbs) of SC IN and MN (laminae II-IV and IX by Rexed) under the influence of PRP-1 (20 - 50 μg/100 g, i/m) and CV at low (3.4 mg/100 g, LD_{50}) and high (6.8 mg/100 g, LD_{100}) doses were recorded. The analysis of results was carried out with a special program, mentioned above in this chapter.

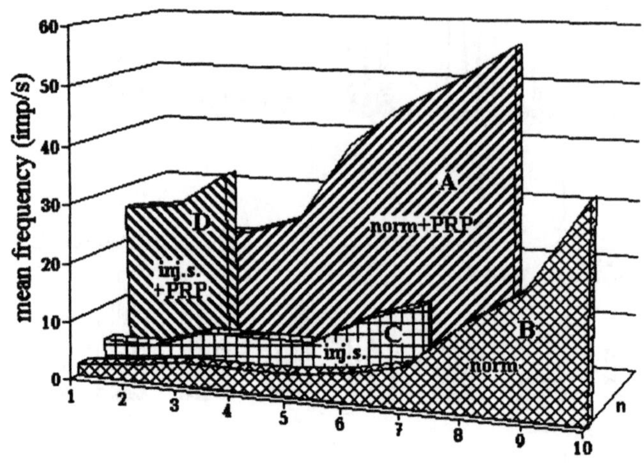

Figure 49. Summarized histograms of BA average frequencies of SC neurons in normal state (A), under the influence of PRP-1 (B), on injured side (C) and on injured side following PRP-1 injection (D). Ordinate, average frequency (imp/sec); abscissa, number of cells (n).

The influence of CV on impulse background and evoked activity with and without PRP-1 treatment was evaluated using the data obtained with 318 single SC MN and IN. Fig. 50 presents a diagram of the relation of mean frequencies of single SC IN and MN BA under separate action of PRP-1 and CV (*B*) and their combined action (*A, C*). In *A* the data of three IN activities are shown: the reduction of BA of IN at the administration of a high dose of CV with subsequent recovery of mean frequency till normal value after PRP-1 injection (*a*, depth 700 μm); amplification of BA at low dose of CV (*b*, depth 600 μm), and the combined effect of CV and PRP-1 (*c*, depth 700 μm).

CV administration leads to a sharp decrease in IN BA down to its elimination; following injection of PRP-1 the mean frequency of IN BA recovers till normal state; repeated administration of CV under PRP-1 action does not result in reduction of IN BA. In *B*, the action of PRP-1 alone on the frequency of BA of two SC MN is shown: in a silent MN BA is registered in response to PRP-1 (20 min after administration, *a*, depth 1200 μm); in the case of another background active MN sharply rise of BA baseline the under the influence of PRP-1 takes place (*b*, depth 1600 μm). In *C* the combined influence of CV and PRP-1 on the alteration of mean frequencies of three MN BA is demonstrated: CV results in significant reduction in mean frequency of BA of MN; administration of PRP-1 restores the reduced BA till normal state (*a*, depth 1200 μm); preliminary administration of PRP-1 leads to an increase in BA of MN, subsequent administration of CV under action of PRP-1 does not negatively affects its frequency (*b*, depth 1400 μm); *c* (depth 1600 μm) reflects the protective effect of PRP-1 that is similar to previously described effect for IN: after brief elevation of MN BA, CV induces a sharp decrease in BA; PRP-1 protects against the following reduction in mean frequency; at the same time, repeated injection of CV is not accompanied by an initial inhibitory effect (*c*).

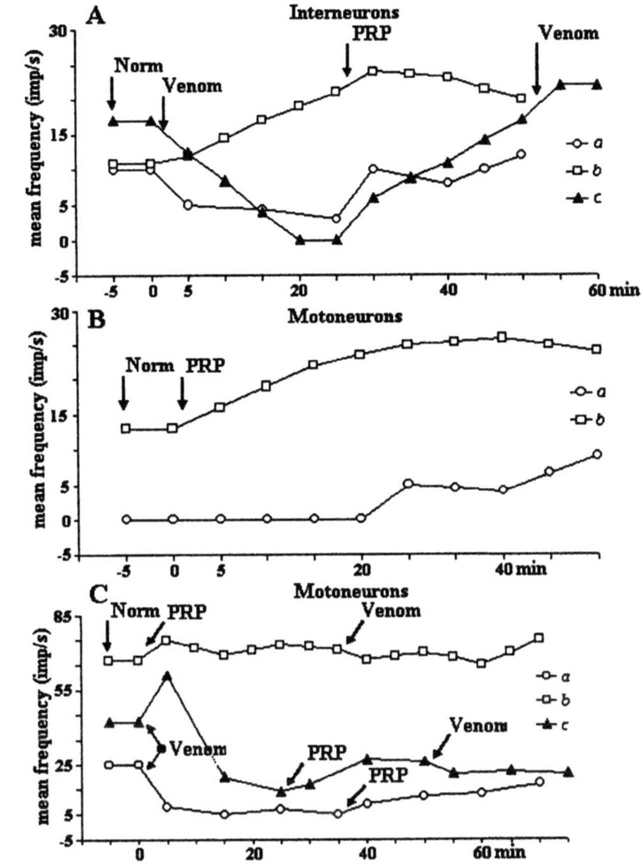

Figure 50. Changes in single SC IN and MN BA (mean frequencies) produced by PRP-1 and CV separate (B) and combined (A, C) influence. A - three IN: *a* (depth 700 μm)- preliminary CV injection (high dose - 6.8 mg/100 g, LD_{100}) and subsequent PRP-1 administration; *b* (depth 600 μm) - CV (low dose - 3.4 mg/100 g, LD_{50}) and PRP-1; *c* (depth 700 μm) - CV plus PRP-1 and repeated CV injection. B: separate action of PRP-1 on the activity of two SC MN; *a* - silent (depth 1200 μm) and *b* - background active (depth 1600 μm) MN. C - three MN: *a* (depth 1200 μm) - preliminary CV and subsequent PRP-1 administration; *b* (depth 1400 μm) - preliminary PRP-1 and subsequent CV administration; *c* (depth 1600 μm) - preliminary CV, subsequent PRP-1: synchronous short-time primary response is observed accompanied by early (with 56 ms) and repeated CV administration. Straight arrow indicates the time of injection and refers to all curves, oblique arrow denotes the same but refers only to a given curve.

On the whole, in all cases with change of mean frequencies of both IN and MN BA the stimulating effect of PRP-1 was revealed. Against a background of PRP-1 (before and after injection of high doses of CV) the typical inhibitory effect was either not detected or taken off. The combination of the facilitating action of low doses of CV and the subsequent facilitating effect of PRP-1 led to the summation of these two effects.

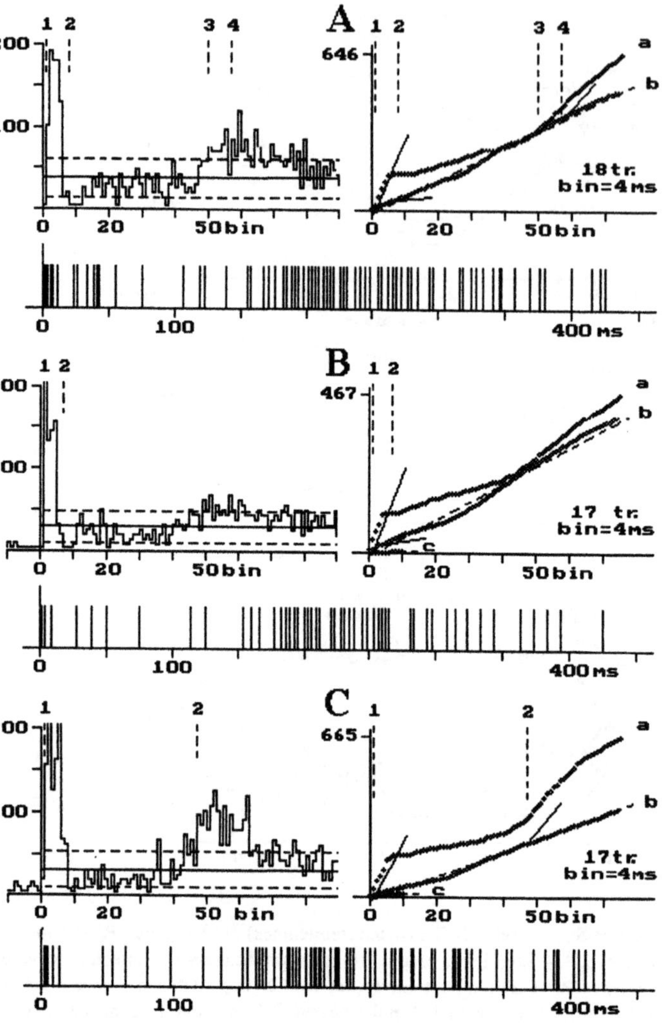

Figure 51. PRP-1 action on impulse flow of single SC MN (depth 1400 μm) at extensor G nerve stimulation after CV administration. A - in normal state (control, non-treated), B - after 25 min of CV and C – after 30 min of PRP-1 administration. Here and on following Figs. 52 - 63: at the left - PETH of interimpulse intervals of a neuron for recorded trials, at the right - diagram of its CC; below - unfolded picture of neuronal activity in selected interval (400 ms) after stimulation. PETH ordinate: percent of spikes number; PETH abscissa: succession of bins; PETH- and CC-numbered dotted lines demonstrate the beginning of excitation or inhibition. CC ordinate: number of spikes; CC abscissa: succession of bins. *a* - CC of poststimulus histogram. *b* – same for tonic activity with its approximated line (dotted). *c* - ditto (to *b*), that and *b*, but for BA. Short segment at angle of approximating line (*b*) indicates the onset of histogram fragment, deviating from M by ± 2SD (the slope of line segment more (less) than the line slope (*b*) means excitation (inhibition) of a neuron). After CV administration, the depression of MN activity is noticed (B) in comparison with normal state (A). Late secondary response (starting with 45 - 50 bins) is also significantly decreased (B). In C – increase in MN activity, absence of depression (with 7 - 8 bins); late secondary response is sufficiently expressed (with 44 - 62 bins).

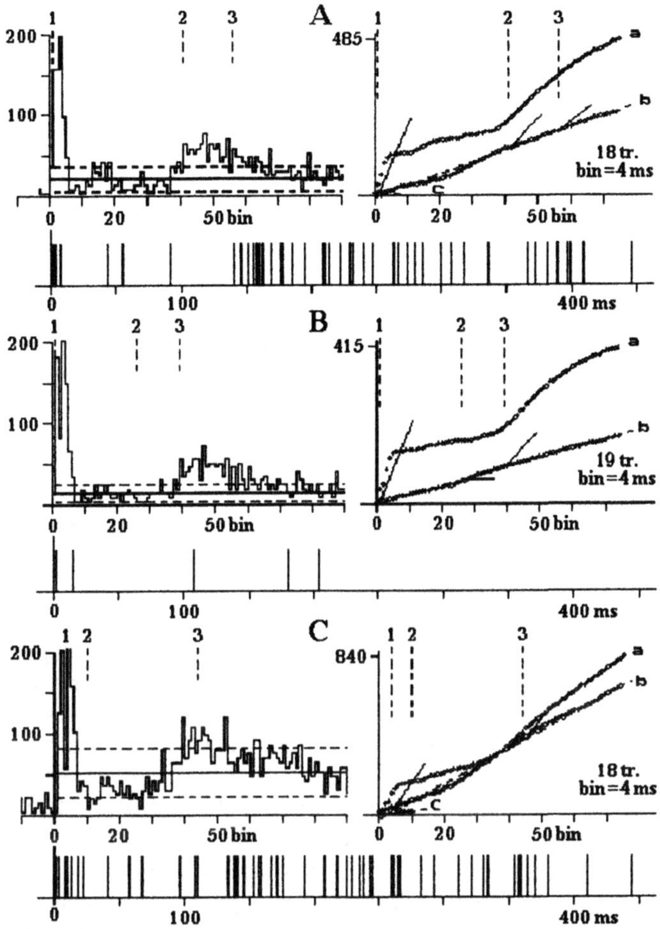

Figure 52. Effect of PRP-1 on SC single MN (depth 1400 μm) impulse flow under flexor P nerve stimulation after preliminary CV administration. A - in normal state (control, non-treated), B - 15 min of CV, and C - 25 min of PRP-1 action. In B –expressed depression at the level of 25 bins is noted. In C – elevation of MN activity (with 40 bins) along with increase in secondary early and late responses are observed.

In Figs. 51-55 the results of statistical analysis of spinal MN and IN spike activity upon the stimulation of corresponding nerves under normal conditions and after separate and combined action of PRP-1 and CV are presented. Each left figure demonstrates PETH of interimpulse intervals of a neuron for recorded trials (in multiple trials in the same order of the effects studied). The right-side diagrams show their cumulative curves (CC). In Fig. 51-55 (below) the neuronal activity in selected interval (400 ms) after stimulation is supplementary unfolded.

In Fig. 51 the alterations of spike activity of single MN (depth 1400 μm) responding to the stimulation of extensor G nerve in normal state (*A*), after CV action (*B*) and subsequent PRP-1 administration (*C*) are demonstrated. As evident from the figure, the poststimulus activity is characterized by synchronous short-time primary response, accompanied by early (within 56 ms, $p < 0.031$) and late (within 164 ms, $p < 0.001$)

secondary responses (multiple discharges of lesser and greater duration and, respectively, lesser and greater expressed) (*A*). After CV administration (*B*) the inhibition of tonic activity (M = 4.6, SD = 2.2) in comparison with normal state (*A*) (M = 6.5, SD = 2.8) is observed. Secondary late response is also significantly weakened (*B*). PRP-1 effect is manifested by the appearance of background discharges and restoration of secondary late response (starting with 180 - 200 ms) (*C*). After PRP-1 administration the tonic activity increases (M = 5.3, SD = 2.6); the inhibition after primary response (within 28 - 32 ms) occurring both in normal state and after CV is taken off; the secondary late response again significantly elevates (*C*).

In Fig. 52 analogous peristimulus effects of CV and PRP-1 on spike activity of MN (depth 1400 μm) responding to the stimulation of flexor nerve P is shown. After CV administration (*B*), a reduction in tonic activity developed (M = 2.6; SD = 1.3), in comparison with normal state (*A*) (M = 3.7; SD = 1.8), including secondary early and late responses. PRP-1 administration (*C*) was accompanied by substantial increase in tonic activity (M = 9.1; SD = 3.6) exceeding the activity of normal state. The reinforcement of secondary responses, predominantly of the late one (with 14 and 40 bins) also took place (*C*). Fig. 53 demonstrates the PRP-1 effect on SC single MN (depth 1700 μm) impulse flow, induced by mixed nerve (n. tibialis) stimulation during the action of CV (*B*), after PRP-1 injection (*C*), and after repeated injection of CV (*D*), in comparison with normal conditions (*A*). The lack of BA and tonic activity after CV injection is clearly seen in ***B***. PRP-1 administration is accompanied by the appearance of background and tonic activities with significant increase in the latter (*C*). After the primary response, the inhibition (with 32 ms) and intense secondary early and late responses (with 56 and 160 ms, respectively) develop (*C*). Repeated injection of CV induces sharp activation of MN (*D*). Significant prevalence of tonic activity (M = 15.1; SD = 6.0) over BA (M = 3.7; SD = 2.2) is observed. Respectively, an increase in secondary early and late responses (with 56 and 184 ms) is notable (*D*).

Fig. 54 presents the effect of separate and combined influence of CV and PRP-1 on impulse flows of SC single IN (depth 900 μm), in response to stimulation of mixed nerve (n. tibialis) after CV injection (*B*), after PRP-1 injection (*C*), and repeated administration of CV (*D*). After CV administration (*B*) the poststimulus activity sharply decreases (total number of spikes in normal state is 644, after CV treatment, 312) which tells on the tonic activity (M = 1.4, SD = 0.5) in comparison with the activity in normal conditions (*A*) (M = 2.6, SD = 1.3). The secondary excitatory effect is completely taken off (after 80 ms) (*B*). PRP-1 partly restores the IN activity (464 spikes); the secondary late response begins to be manifested as well (*C*). After repeated administration of CV a restoration of baseline level of activity, including BA and secondary responses, is noticed (*D*).

Fig. 55 demonstrates similar effects of CV and PRP-1 on SC single IN under the stimulation of flexor P nerve (depth 900 μm) after CV action (*B*), after PRP-1 injection (*C*), and repeated CV administration (*D*). As seen in *B*, after CV administration the poststimulus activity decreases (total number of spikes under normal conditions (*A*) is 527, after the venom treatment is 225) as well as tonic activity (from M = 2.6, SD = 1.4 to M = 1.7, SD = 0.8). The inhibition after primary response is strongly expressed (*B*); secondary late response under normal conditions (*A*) (within 160 ms) after CV treatment appears for a short period (12 - 15 ms) (*B*) instead of 40 - 50 ms in normal state (*A*). As seen in *C*, PRP-1 administration leads to increase in total activity (1003 spikes compared to 225 spikes after CV administration). IN starts to display high activity in the initial inhibitory period which develops both in normal state (*A*) and after CV injection (*C*). In *D*, the spike activity after repeated administration of CV starts to approach the normal state level.

Thus, we studied separate and combined effects of the new hypothalamic neurohormone, PRP-1, in acute neuronal injury in the SC. Neuronal injury was induced by CV intoxication. In the experiments on intact spinal rats and rats with envenomation followed by PRP-1 treatment, changes in background and evoked impulse flows of single IN and MN of lumbal region (II - IV and IX laminae by Rexed) were recorded. Evoked activity was elicited by stimulation of flexor (P), extensor (G), and mixed (n. tibialis) nerves of hind limbs.

According to the results of the present study, the facilitating action of PRP-1 on the above mentioned neurons was revealed. High doses of the CV caused a depressive effect. In this case PRP-1 administration both before and after the intoxication restored the electrical activity to the baseline level and higher. Small doses induced a facilitating effect. In several cases the repeated administration of the CV restrained the expressed effect of the PRP-1 till the baseline level (for example, the inhibition of activity preceding secondary late response). In addition, higher susceptibility of secondary early and late responses of different types of spinal neurons to chemical influence was detected. The suppression of CV-induced pathological activity with PRP-1 is of interest.

The only antivenom available in the US for the treatment of pit viper envenomation is Antivenin (Crotalidae) Polyvalent (ACP). However, short half-life of the antivenom Fab-molecules (CroFab, FabAV and ACT) can promote the recurrence of the venom effects. Moreover, Fab preparations may develop acute or delayed allergic reactions, though injecting the preparation quickly, and thus shortening the period of antivenom administration, stabilize the antivenom in extreme conditions of heat and handling[218]. The preparation PRP-1 does not manifest any side effects. PRP-1 is sufficiently stable *in vivo* and effective being injected both before and after intoxication (even when more than one lethal dose of venom is administered). Thus, our data provide evidence that PRP-1 is a universal and efficient antivenom agent.

Numerous literature data have been accumulated on biochemical mechanisms of nerve growth factor (NGF) treatment of different neuronal injuries[218-232] and on the role of phospholipase A_2 (PLA_2) in the pathogenesis of snake venoms[217,233-240]. The neuroprotective effect against radical–induced toxicity in cultured spinal MN was achieved by the use of cGMP[241].

In view of the neuroprotective role of PRP-1, the presence of specific receptors to PRP-1 on neurons and glia needs to be examined. Possible caspase participation and the multiple mechanisms of PRP-1 action (immunological, biochemical, and genetical processes), should be elucidated. In accord with our results, PRP-1 inhibits PLA_2 activity, thus contributing to the protection against membrane oxidative damage[242-243]. The action of venoms of different snake species on the activity of this enzyme was reported, and different neuroprotectors were tested: steroids, (in particular methylprednisolone, DEX), interleukins, melatonin, calpain inhibitors, cGMP[241]. Venoms affecting cholinergic neurons were also reported.

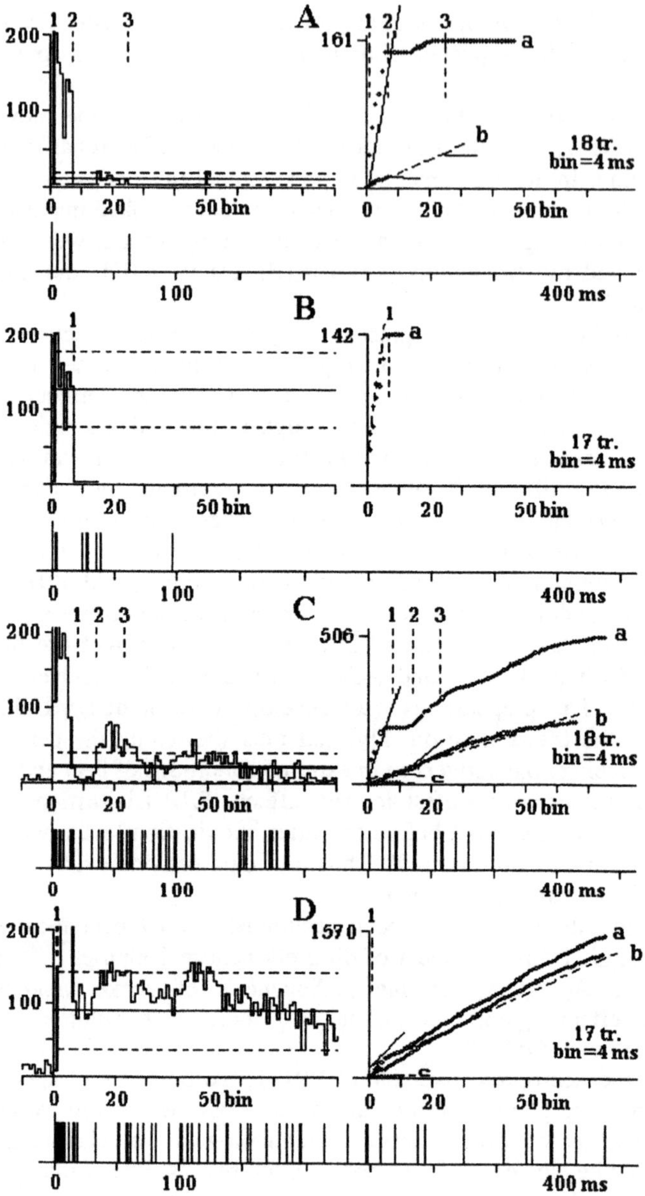

Figure 53. Effect of PRP-1 on SC single MN (depth 1700 μm) impulse flow. Mixed nerve (n. tibialis) stimulation, preliminary and repeated administration of CV. In normal state (A, control, non-treated), 15 min after CV (B), 45 min after PRP-1 (C), and 2 min after repeated injection of CV (D). In B lack of activity is seen. In C – appearance of background and tonic activity with significant predomination of the latter is observed. After primary response, the inhibition (with 8 bin) and expressed secondary early and late responses developed (with 14 and 40 bins, respectively). In D – sharp increase in MN activity after repeated administration of CV. Significant prevalence of tonic activity over BA is noticeable; elevation of secondary early and late responses takes place (with 14 and 46 bins, respectively) (D).

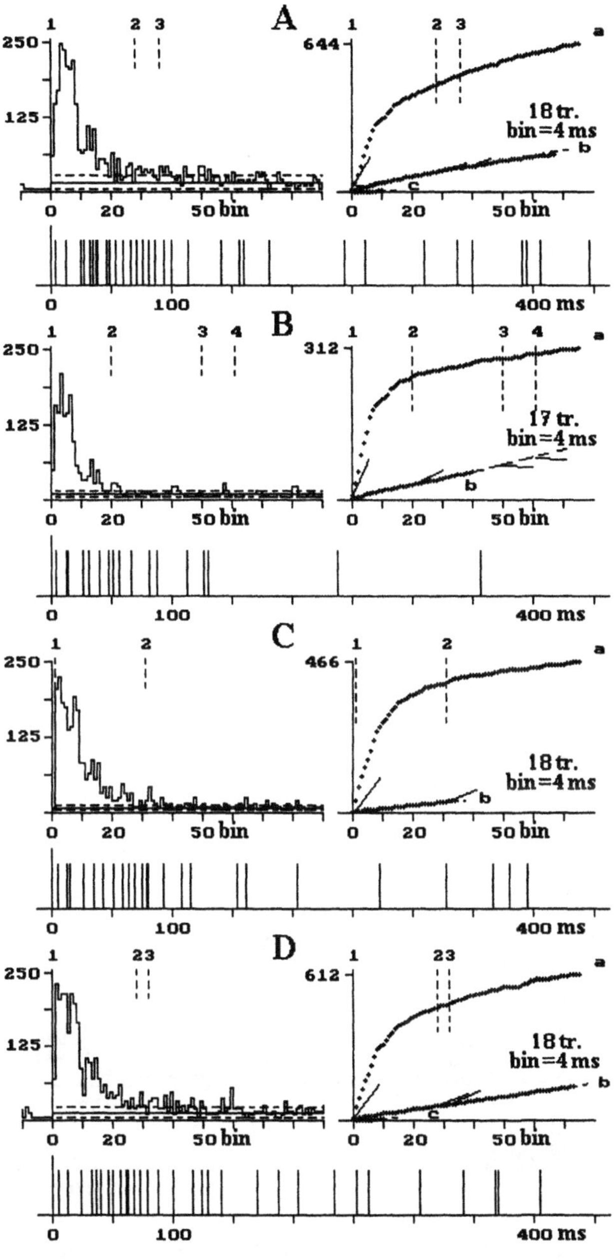

Figure 54. Effect of separate and combined influence of CV and PRP-1 on impulse flows of SC single IN (depth 900 μm) responding to stimulation of mixed nerve (n. tibialis). In normal state (A, control, non-treated), after 20 min of CV (B), 45 min of PRP-1 (C) and 15 min of CV repeated administration (D). In B - poststimulus activity sharply decreases that tells upon tonic activity in comparison with normal state (A). Secondary excitatory effect disappears (with 20 bins). In C - PRP-1 partly restores the activity; secondary late response begins to be manifested as well. In D - after CV repeated administration a restoration of baseline level of activity, including BA and secondary responses, is noticed.

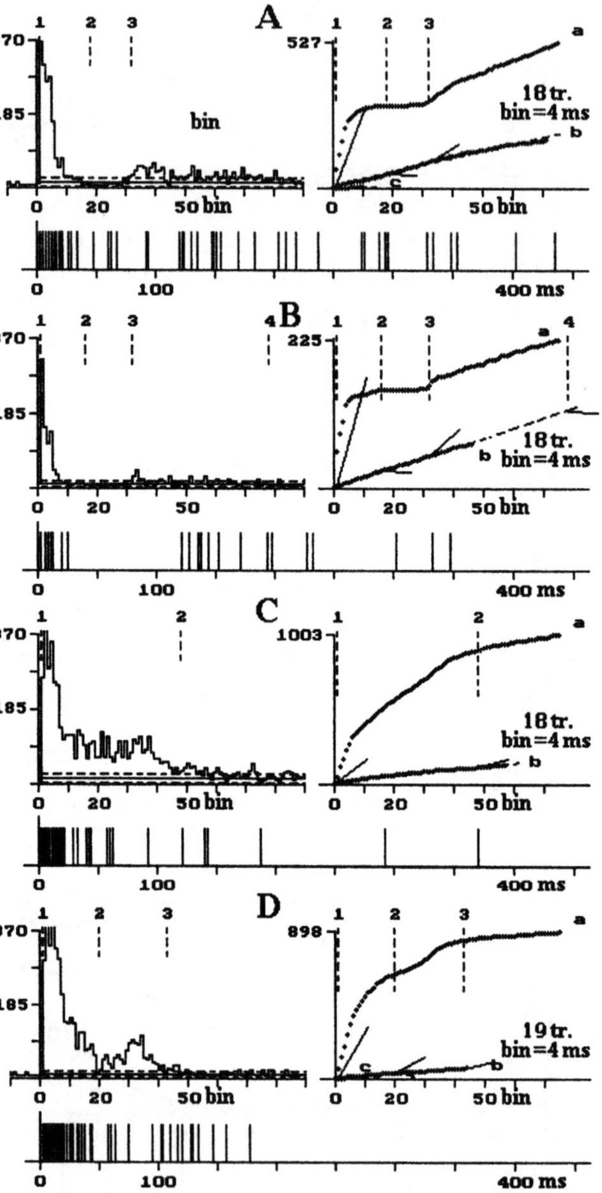

Figure 55. Effect of CV and PRP-1 on SC single IN (depth 900 μm) upon stimulation of flexor nerve - P. In normal state (A, control, non-treated), 10 min after CV administration (B), 45 min after PRP-1 administration (C), and 15 min after CV repeated administration (D). In B - poststimulus activity decreases and the tonic activity falls. Inhibition after primary response is highly expressed (B). Secondary late response (with 40 bins) after CV administration appears for a short period (3 - 5 bins) (B) instead of 10 - 12.5 bins in normal state (A). In C - PRP-1 administration leads to a significant increase in total activity. IN display high activity in the initial inhibitory period occurring both in normal state (A) and after CV injection (C). In D - spike activity curve after repeated CV administration approaches to that of normal state (A) by shape, remaining still higher.

To uncover common mechanisms of venom action on neurons, the following snakes venoms and toxins have been investigated: Naja haje venom[233], Chinese cobra venom[244], sarafotoxin[245], venom of Vipera ammodytes[240], Dendroaspis angusticeps venom[246-250], beta-bungaratoxin[251], Naja naja astra venom, cobrotoxin, ammodytoxin C, Bothrops lanceolathus (Fer de lance) venom, Bothrops jararaka venom. The evaluation of the efficacy of different neuroprotectors against venom injury allows us to state that PRP-1 is distinct from other drugs by the dosage used and the effectiveness of recovery. PRP-1 is one of the most efficacious drugs known for the treatment of acute neurodegenerative diseases.

6.2. PROTECTIVE EFFECT OF PRP-1 AGAINST TRAUMA-INDUCED CENTRAL AND PERIPHERAL NEURONAL INJURY

The purpose of this investigation was to study the effect of PRP-1 on trauma-damaged SC and to perform morpho-histological analysis of recovery phases and growth of neuronal fibers in SC, and also of glia proliferation and maintenance of MN integrity and survival.

The experiments were conducted on 87 adult male and female rats (Wistar, 200-300g) with chronic neurodegeneration, including SC lateral hemisection (18 rats) and I-nerve section (11 rats). Rats were subjected to SC hemisection at L_2 - L_3 level under Nembutal anesthesia. In other group of rats, I nerve cuts were made in the upper third of hind limb. The animals with hemisection received daily injections of PRP-1 (20 mkg/100 g, i/m) from first day of the postoperation period, for 3 weeks, or injections started one week later and lasted for 2 weeks. Control animals received injections of saline.

This model of trauma allows a precise differentiation between damaged (below section) and intact control (symmetric opposite) regions.

Evoked responses to stimulation of the flexor – P, extensor – G, and mixed – I nerves of hind limb were recorded from single IN and MN of SC of the hemisected animals (laminae II-IV and VII-IX by Rexed) after 3 weeks, with and without PRP-1 treatment.

Another group of animals (I nerve cut) received PRP-1 (10 mkg/100 g i/m) or saline injections. In this case evoked spike activities of single IN and MN of SC (laminae II-IV and VII-IX by Rexed) in response to stimulation of the nerve were recorded (Galoyan et al., in press).

The effect of SC trauma (hemisection, 54 rats used, and axotomy, 33 rats) on MN and IN impulse evoked activity, with and without PRP-1 treatment, was studied. Data were obtained from 276 (hemisection) and 357 (axotomy) individual MN and IN in SC-injured rats, including 18 animals (276 MN and IN) with preliminary (3 - 4 weeks before) SC hemisections and 11 animals (228 MN and IN) with previous I nerve cuts.

In Figs. 56 – 59 the changes in IN (Figs. 56, 57) and MN (Figs. 58 - 59) response under stimulation of extensor (G), flexor (P), and mixed (I) nerves of hind limbs on the injured and uninjured sides of SC, under conditions of hemisection, with and without PRP-1 treatment, are shown.

On days 21 - 22 after hemisection in control (shown in Fig. 56) significant decrease of IN pre- and postsynaptic activity on injury side below the place of the cut under G (*A*) and P (*B*) nerves stimulation are already noticeable; background activity (BA) is vanishing and after stimulation reveals a define tonic activity.

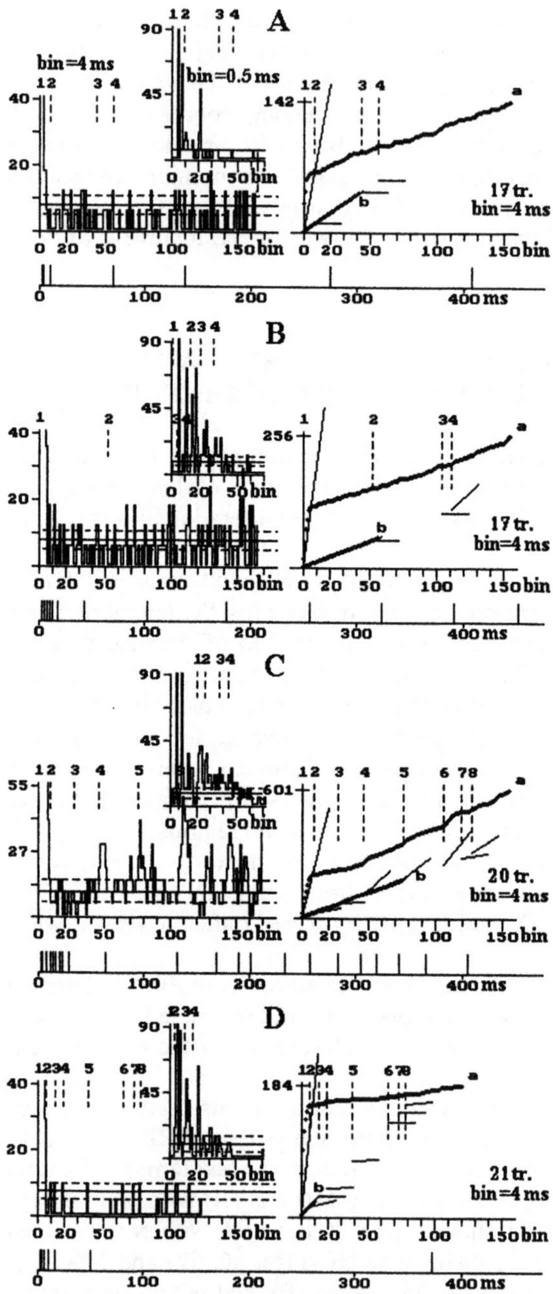

Figure 56. Spike activity of single SC IN on stimulation G (A, C) and P (B, D) nerves 22 (A, B) and 21 (C, D) days after SC hemisection without PRP-1 administration. A, B and C, D - 2 IN (depth 750 and 500 μm, respectively) beyond 1.0 (A, B) and 0.5 (C, D) mm below the cut on injury side (A, B) and symmetric (non-injured) side of SC (C, D).

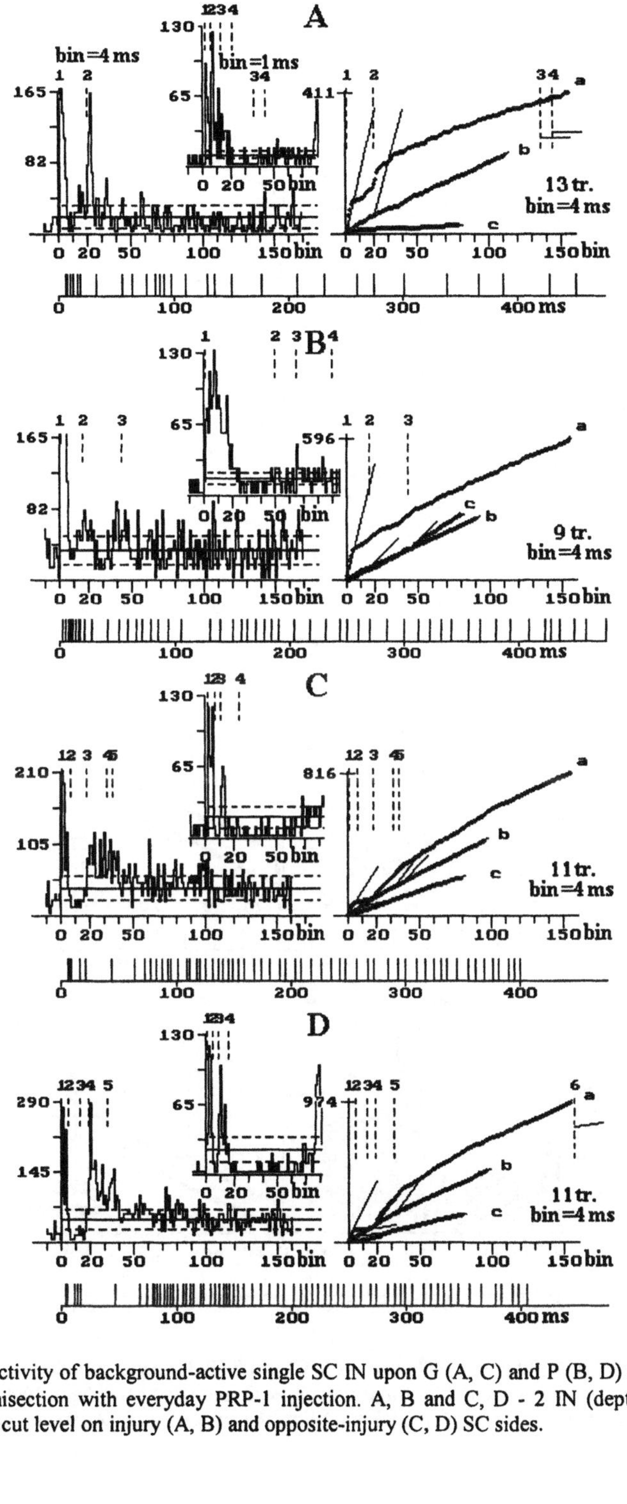

Figure 57. Spike activity of background-active single SC IN upon G (A, C) and P (B, D) nerve stimulation 22 days after SC hemisection with everyday PRP-1 injection. A, B and C, D - 2 IN (depth 450 and 200 μm, respectively) at the cut level on injury (A, B) and opposite-injury (C, D) SC sides.

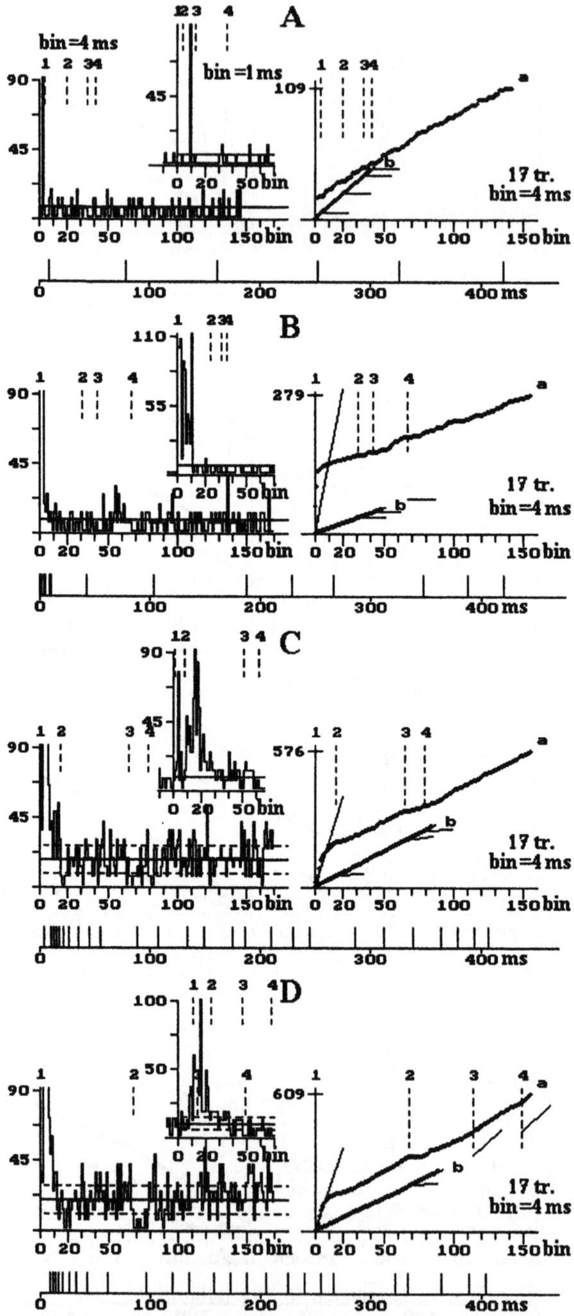

Figure 58. Spike activity of single SC MN upon G (A, C) and P (B, D) nerve stimulation, 23 (A, B) and 22 (C, D) days after SC hemisection without PRP-1 injection. A, B, C, and D - 3 MN (depth 1000, 1000 and 1100 μm, respectively) between 1.0 (A, B) and 2.0 (C, D) mm below the cut, on injury (A, B) and opposite-injury (C, D) SC sides.

On the uninjured side with stimulation of G and P nerves (*C* and *D*, respectively) a marked decrease in the activity also takes place, but with a diffuse increase in its late components (sporadic splashes).

The Fig. 57 illustrates SC IN activity after G and Pnerve stimulation on the injury side 22 days after SC hemisection (on the section level), in the conditions of PRP-1 administration every day from the first postoperation day.

Fig. 58 demonstrates a common decrease in MN activity 23 days (*A*, *B*) and 22 (*C*, *D*) days after hemisection below the place of the cut in control (without PRP-1 injection).

Under G nerve stimulation, a short primary response at 10 ms is noticed on the injury side which is accompanied by a weak tonic activation (*A*); besides, on the same depth on the injury side the primary response of the same nerve is recorded at 3 - 5 ms, then subsequent primary late response, at 10 - 30 ms, accompanied by tonic activation (*C*).

Under P nerve stimulation, the early primary response was registered on the injury side within 3 to 10 ms (*B*), and the late primary response appeared on the uninjured side within 10 to 25 ms (*D*), both accompanied by tonic activity.

Finally, Fig. 59 demonstrates the type of MN response on the level of the section 22 days after SC hemisection, with PRP-1 daily administration starting from the first day. Well-defined and more prolonged primary and secondary components of response are observed on the injury side under G and P nerve stimulation (*A*, *B*). Furthermore, under P nerve stimulation the late primary component (*B*), inherent to the response of the G nerve stimulation, is noticeable.

Fig. 60 and Fig. 61 present the trials of IN and MN activities (axotomy had been performed 3 weeks before on the injury side), after stimulation of mixed (I), extensor (G), and flexor (P) nerves without administration of PRP-1.

Fig. 60 demonstrates the effects of stimulation of two IN (*A* and *B* - *D*, respectively) on the injury side. Under stimulation of ipsilateral I (Ii) nerve primary early response occurs in 8 ms, transition to primary late response takes places in 15 ms, after that marked inhibition within 50 ms proceeds, and the formation of secondary early response accompanied by repeated inhibition takes place in the interval from 160 to 200 ms (*A*). This IN did not respond on G and P nerve stimulation on the same side.

The other neuron under stimulation the Ii nerve on the injury side after a short primary response at 8 ms reacted by inhibition during 200 ms of poststimulus period, then the recovery to the level of BA took place (*B*).

In absence of IN response on ipsilateral G (Gi) and ipsilateral P (Pi) nerve stimulation, upon contralateral G (Gc) stimulation after a period of activity inhibition from 10 to 70 ms and default of primary response, early secondary response appeared in the period from 80 to 120 ms, after that remote inhibition occurred accompanied by weak but noticeable activation (*C*).

In the above cases the approach at the level of phasic and tonic activities would be of interest.

Finally, in *D* is seen that under stimulation of contralateral P (Pc) nerve, the inhibition in the interval from 25 to 70 ms occurs accompanied by shifts in poststimulus activity, the result analogous to that shown in *C*. The neuron does not respond to the stimulation of Gi and Pi, i.e. stimulation on the injury side. Long latency of contralateral responses and absence of ipsilateral ones should be noted.

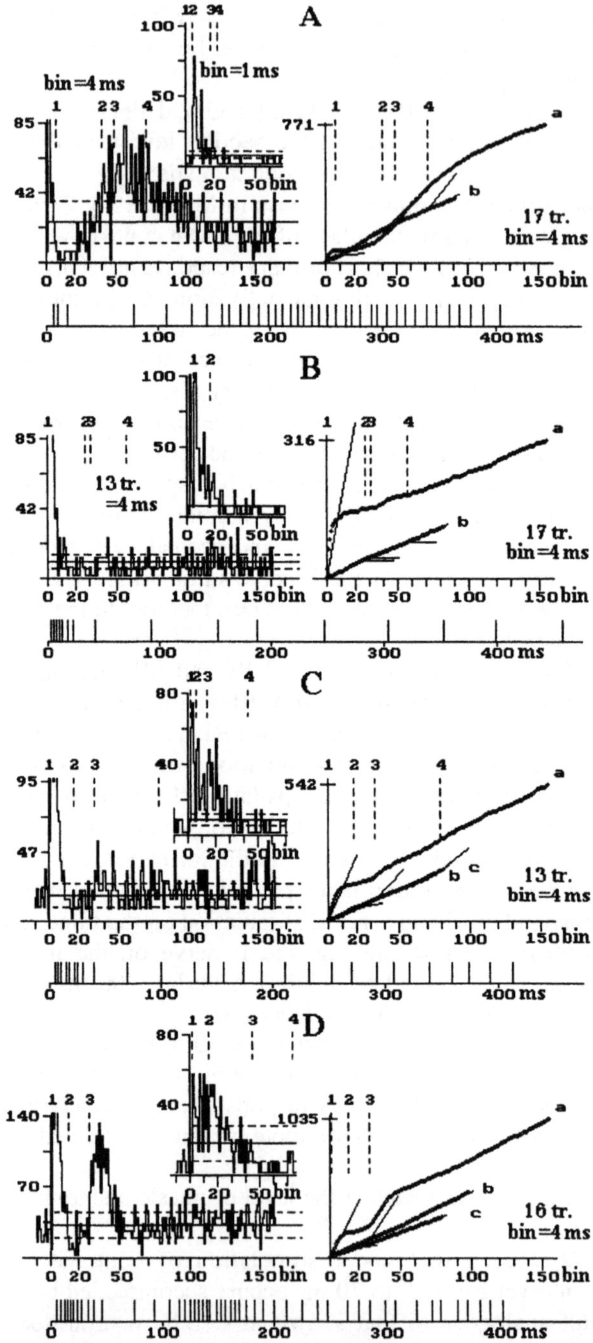

Figure 59. Spike activity of background active single SC MN upon G (A, C) and P (B, D) nerve stimulation, 22 days after SC hemisection with everyday PRP-1 injection. A, B and C, D - 2 MN (depth 1200 μm) between 0.2 mm below- and at the cut level on injury (A, B) and opposite-injury (C, D) sides of SC.

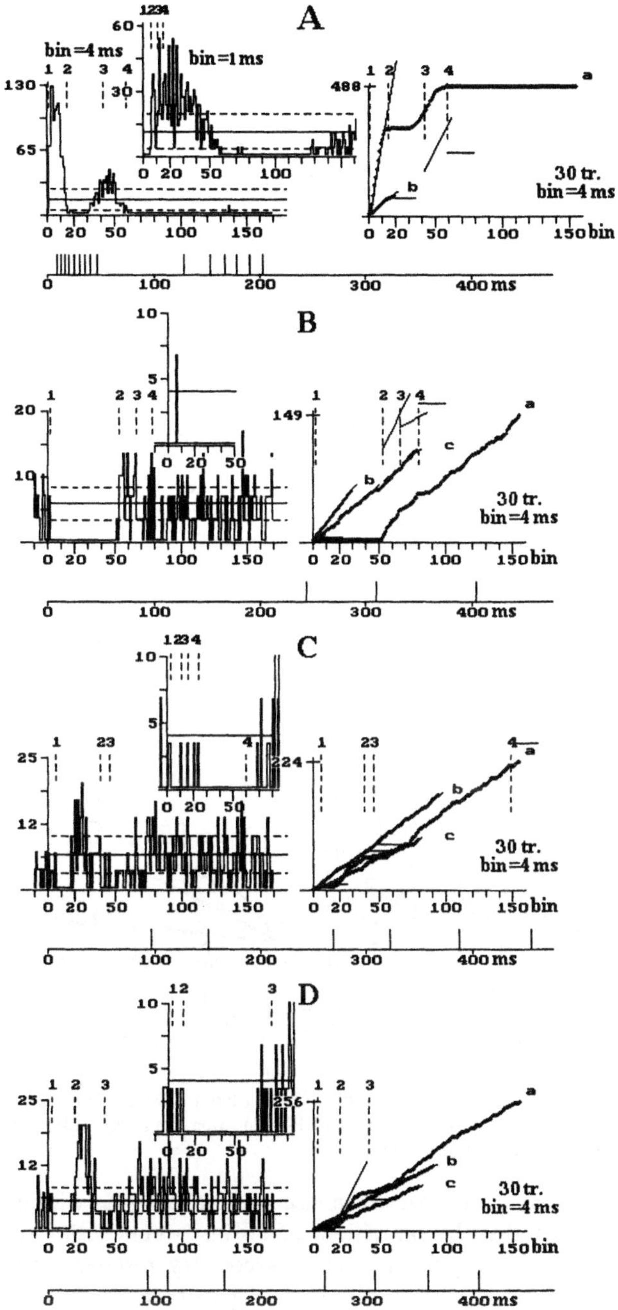

Figure 60. Spike activity of two background-active single SC IN upon ipsilateral mixed I (Ii) (A, B), contralateral extensor - Gc (C), and flexor - Pc (D) nerve stimulation 3 weeks after axotomy without PRP-1 administration, on injury side of SC. A and B-D – 2 IN (depth 850 и 750 μm, respectively); IN in A did not respond on the stimulation of above mentioned ipsilateral extensor and flexor nerves.

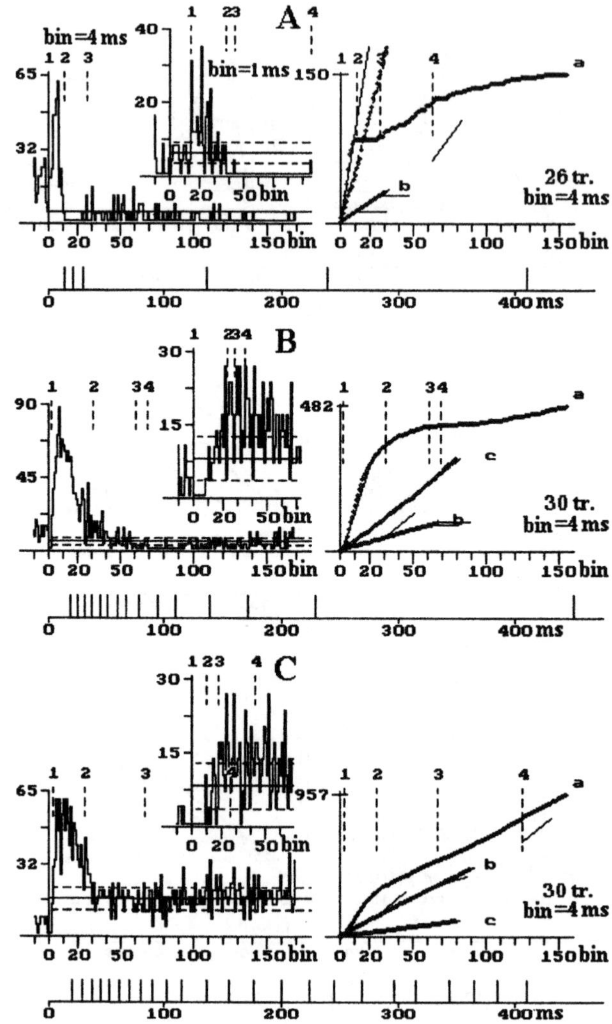

Figure 61. Spike activity of background-active single SC MN (depth 1400 μm) upon Ii (A), Gc (B), and Pc (C) nerve stimulation 3 weeks after axotomy, without PRP-1 injection on injury side; MN did not respond to ipsilateral extensor and flexor nerve stimulation.

Fig. 61 demonstrates the primary response on Ii stimulation within the interval from 12 to 18 - 20 ms; after that the inhibition till 80 ms occurs followed by a very weak increase in the activity during the period of secondary response, after which its fast decrease to a level below BA takes place (*A*). Upon stimulation of Gc a pronounced activity was observed for a long period from 10 to 90-100 ms; afterwards, the activity decreased below the background level (*C*), as on *B*, but the difference was in the maintenance of tonic activity for a long time over BA level (*C*). The neuron did not respond on Gi and Pi nerve stimulation.

Thus, as was expected, in weeks after the injury MN and IN did not respond to the stimulation of extensor (G) and flexor (P) nerves in distal stump on the injury side in the control (without PRP-1) owing to the absence of concretion of separated terminals and atrophy of distal SC part. However, the stimulation of proximal stump of I nerve often initiated the abovementioned activation of SC neuronal elements.

Figs. 62 and 63 demonstrate, by recording of IN and MN single impulse activity, the effects of I, G, and P nerve stimulation on ipsilateral injury side in control and under the action of PRP-1.

In Fig. 62, early primary response on the stimulation of Ii in SC IN on injury side is registered between 8 and 40 ms; it persists less intensely till to 100 - 120 ms, with subsequent gradual decrease to a level of tonic activity exceeding BA (*A*).

The stimulation of Gi gives rise to primary response which starts between 40 and 80 - 90 ms; then tonic activity slightly exceeding the BA level is maintained (*B*).

Under Pi stimulation, the activity is registered till 180 ms, then the inhibition of the activity occurrs up to 200 ms, after which a pronounced response takes place accompanied by activity recovery to the level of tonic activity, close to BA (*C*).

Under stimulation of Gc, the primary response occurs between 8 and 20 ms, then the tonic activity somewhat above BA is recorded (D).

Fig. 63 shows the initiation of SC MN primary response on the injury side on the stimulation of Ii which takes place in the interval from 10 to 20 ms, with subsequent inhibition, which starts at 40 ms, followed by well marked activity between 80 and 240 ms, and accompanied by a decrease in the activity to the level of tonic activity (*A*).

Under Gi stimulation MN begin to produce primary response at 15 ms, the activation lasts up to 75-80 ms, then an increase in the activity is registered, pronounced at 200 ms, with subsequent slow decrease to the level of tonic activity, considerably higher than BA (*B*).

Pi stimulation was followed by the initiation of primary activity between 10 and 20 ms, the next pronounced activity manifestation was observed at 40 ms, after that the inhibition at 40-50 ms followed. An increase in the activity took place between 80 and 300 ms followed by subsequent decrease to tonic activity level, significantly exceeding BA (*C*).

Under Gc stimulation, the primary response occurred between 10 and 25 ms, then prolonged activation lasted till 300 ms, accompanied by a decrease in the activity to the level of tonic activity which also significantly exceeded BA (*D*).

In the case of Pc stimulation, primary response between 8 and 25 ms was revealed, then the activity inhibition was observed in 40-55 ms interval; subsequent splashes of activity between 80 and 250 ms were registered accompanied by its decrease to the level of tonic activity somewhat over BA level (*E*).

Thus, in the experiments with PRP-1, in animals with axotomy, the above mentioned nerve stimulation evoked responses of SC IN and MN were recorded on the injury side. The phenomenon indicated the restoration of IN and MN evoked activity and revealed partial deafferentation, reaffirming the concretion of proximal and distal stumps of transected nerves.

Moreover, the enlargement of distal parts of nerves under study took place, and the recovery of motor activity was observed.

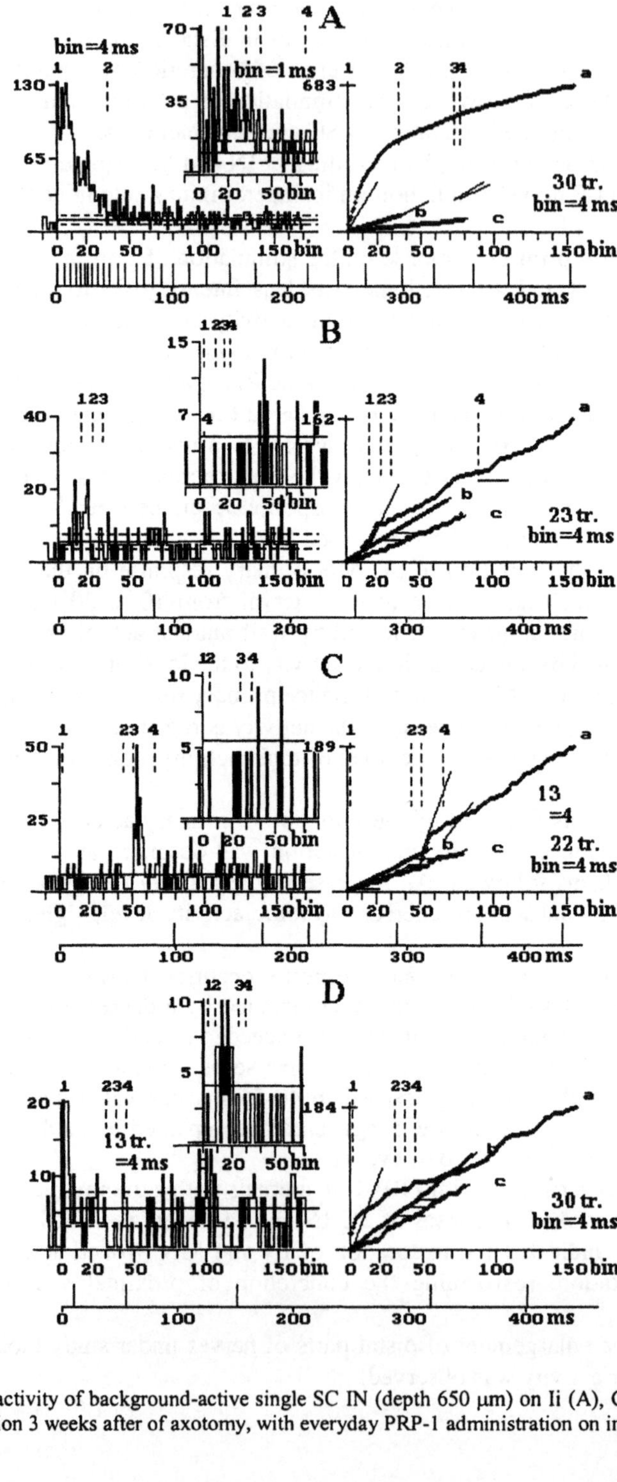

Figure 62. Spike activity of background-active single SC IN (depth 650 μm) on Ii (A), Gi (B), Pi (C), and Gc (D) nerve stimulation 3 weeks after of axotomy, with everyday PRP-1 administration on injury side of SC.

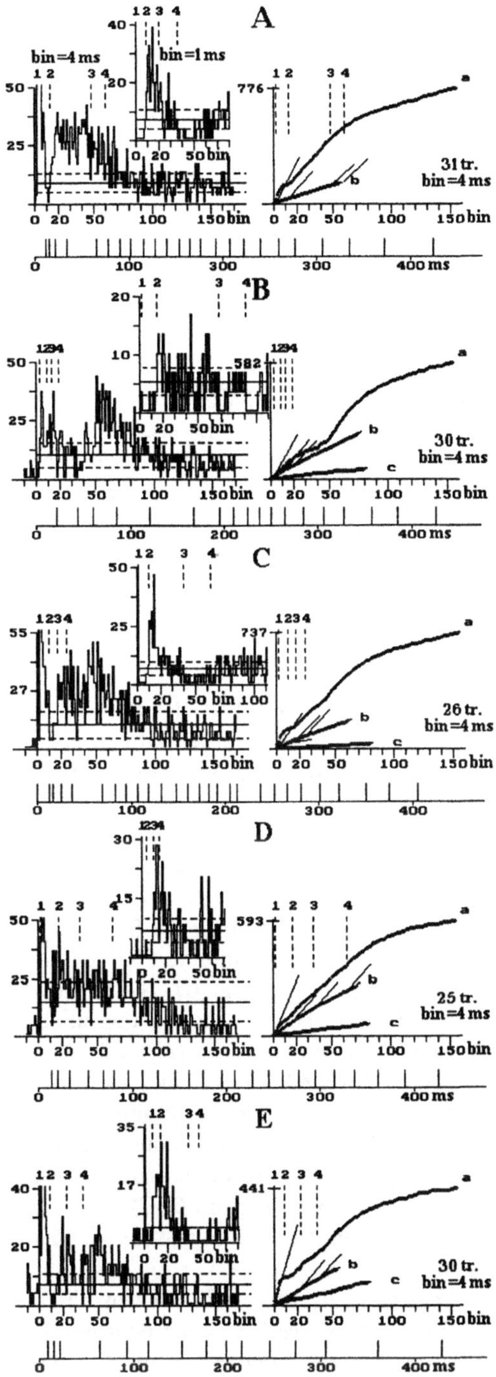

Figure 63. Spike activity of BA single SC MN (depth 1300 μm) upon Ii (A), Gi (B), Pi (C), Gc (D) and Pc (E) nerve stimulation 3 week after axotomy, with everyday PRP-1 injection on injury side of SC.

In experiments described above in this chapter electrophysiological approach was employed. For histochemical examination, the corresponding region of SC was fixed in 10% neutral formalin for 1 month. Frozen horizontal 50 μm-slices were prepared from SC pieces. The slices were placed in $FeCl_3$ solution for 15-20 min, and after careful washing were transferred into incubation mixture prepared according to modified Homori method for detection of acid phosphatase activity[252]. After washing the slices by distilled water they were treated by 3% sodium sulfide (Na_2S) solution and placed in balm. Then free-floating 50-μm sections were cut using a freezing microtome and used for immunohistochemical examination by the Avidin-Biotin Peroxidase Complex (ABC) method[253]. Using an antiserum (1:5,000) raised by Ambrosius[254] against PRP-1 synthetic preparation we studied the localization of PRP-1 in the SC structures of the experimental rats. Slices treated with normal goat serum (NGS, 1:5,000) served as negative control.

Histochemical analysis 3 weeks after the section showed that without PRP-1 injection (control) the injured region of SC was filled with connective tissue forming a scar. Separated nerve fibers were seen above and below the section (Fig. 64*A*).

Daily administration of PRP-1 for 2 weeks, starting one week after hemisection, resulted in restoration of nerve fibers and neuro-glial elements in the damaged region (Fig. 64 *B, C*).

As shown in Fig. 64, *B* and *C*, the growth of the nerve fibers and glial cells into the injured zone takes place. On the surface of SC in the section site some thickening of the dura matter is observed.

Four weeks after administration of PRP-1, starting from the day following hemisection, the damaged zone was filled with neural cell elements. As is obvious from Fig. 65, the post-traumatic zone of SC is completely filled with neurons (*A, B*), nerve fibers, and glial cells (*C, D*).

In the previous studies we observed that in intact rat SC PRP-1 immunoreactivity was associated with glial elements, nerve fibers, and neurons of different size.

In experiments presented in Fig. 66 (*A, B*), no PRP-1-immunoreactive structures were found in the hemisection area 4 weeks after the SC injury. In animals injected with PRP-1 for 4 weeks immediately following injury, PRP-1-immunoreactive nerve fibers and bundles were found (Fig. 66, *C, D*; Fig. 67, *A-D*). In addition, after the PRP-1 administration a gliosis is visible in the hemisection region (Fig. 67, *D*). The nuclei of glial cells display strong PRP-1 immunoreactivity (Fig. 67, *D*).

Besides, the hemisection resulted in an increase in the number of PRP-1-immunoreactive MN (Fig. 67, *A-D*) and their migration into the region of the section (Fig. 67 *A, B*). The hemisection also resulted in changes of MN morphology (Fig. 68*A*). The neurons lose their processes, and their form and size change. In animals injected with PRP-1, the survival of MN was observed (Fig. 68 *B, C*).

The number of PRP-1-immunopositive glial cell nuclei was decreased after complete regeneration of MN (Fig. 68*C*).

In control sections where the antiserum to PRP-1 was replaced by normal goat serum (NGS), there was no staining of the nerve fibers and neurons (Fig. 66 *E, F* and Fig. 67 *E, F*), the fact indicating the specificity of neuronal staining. Glial nuclei, however, stained positively with NGS[212,216].

Morpho-functional experiments revealed the exclusion of difference in background and evoked activity of neurons below section and on intact symmetric side after 2-4-week daily administration of PRP-1.

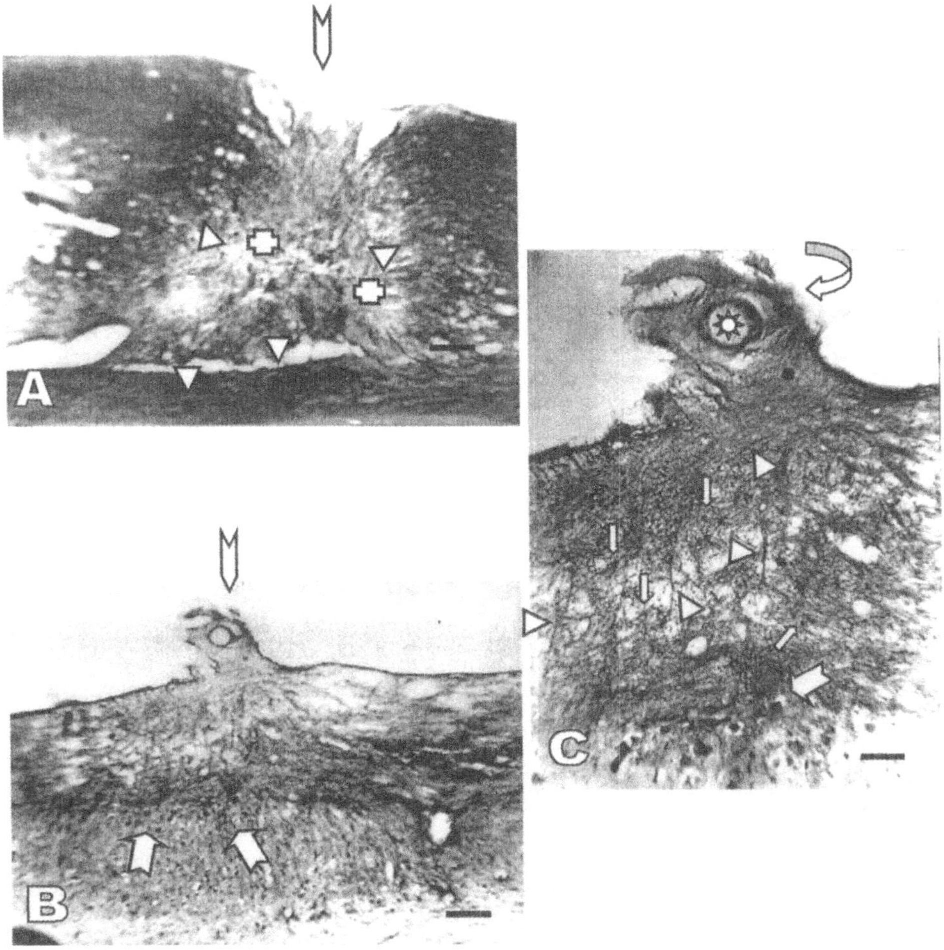

Figure 64. Histochemically stained neuronal structures. A: transection site of SC (chevron). The connective tissue scar (crosses) formed 4 weeks after hemisection; at both right and left sides the cut nerve fibers (triangles) are seen. B, C: daily PRP-1 administered for 2 weeks causes nerve fiber regeneration, survival of MN (large arrow), and increase in gliocyte nuclei (small arrows) in the transected site, from second week after hemisection; fine nerve fibers and transverse section of a blood vessel (sun) are seen in the overgrowth area of dura matter (curved arrow); C: detail of B. Bars: 100 μm (A, B), 50 μm (C).

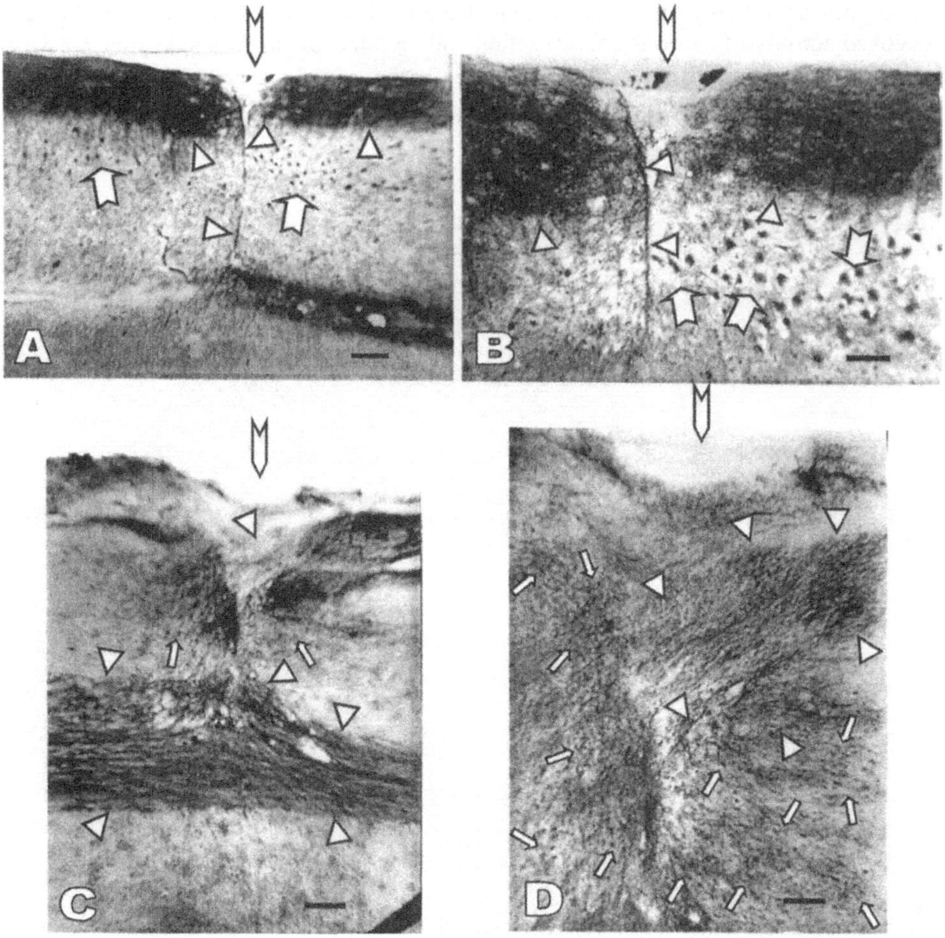

Figure 65. Survival of SC MN and regeneration of nerve fibers after administration of PRP-1 for 4 weeks starting from the first day after hemisection. Transected place is noticed by a chevron. Complete regeneration of transected region with well-observed migration of MN (large arrows) to cutting place (A, B) and (C, D) of regenerated nerve fibers (triangles). C, D: between nerve fibers gliocyte nuclei are visible (small arrows). B: detail of A; D: detail of C. Bars: 100 μm (A, C), 50 μm (B, D).

Under PRP-1 action the inhibition of scar formation, efficient neuroglia proliferation, impairing of white substance growth, protection against cell degeneration, and suppression of tissue stress were discovered in our experiments.

These effects show the perspective of PRP-1 use in clinical practice for the rehabilitation of exotoxic effects and traumas and for the treatment of insults and specific neurodegenerative diseases.

Summarizing, we can conclude, that in the case of injury, the effect of PRP-1 on the recovery and total increase in the activity of neuronal elements of SC below the cut and even at the level of cut results in the maintenance of regeneration and prevention of scar formation, which is proved by the above morphological data.

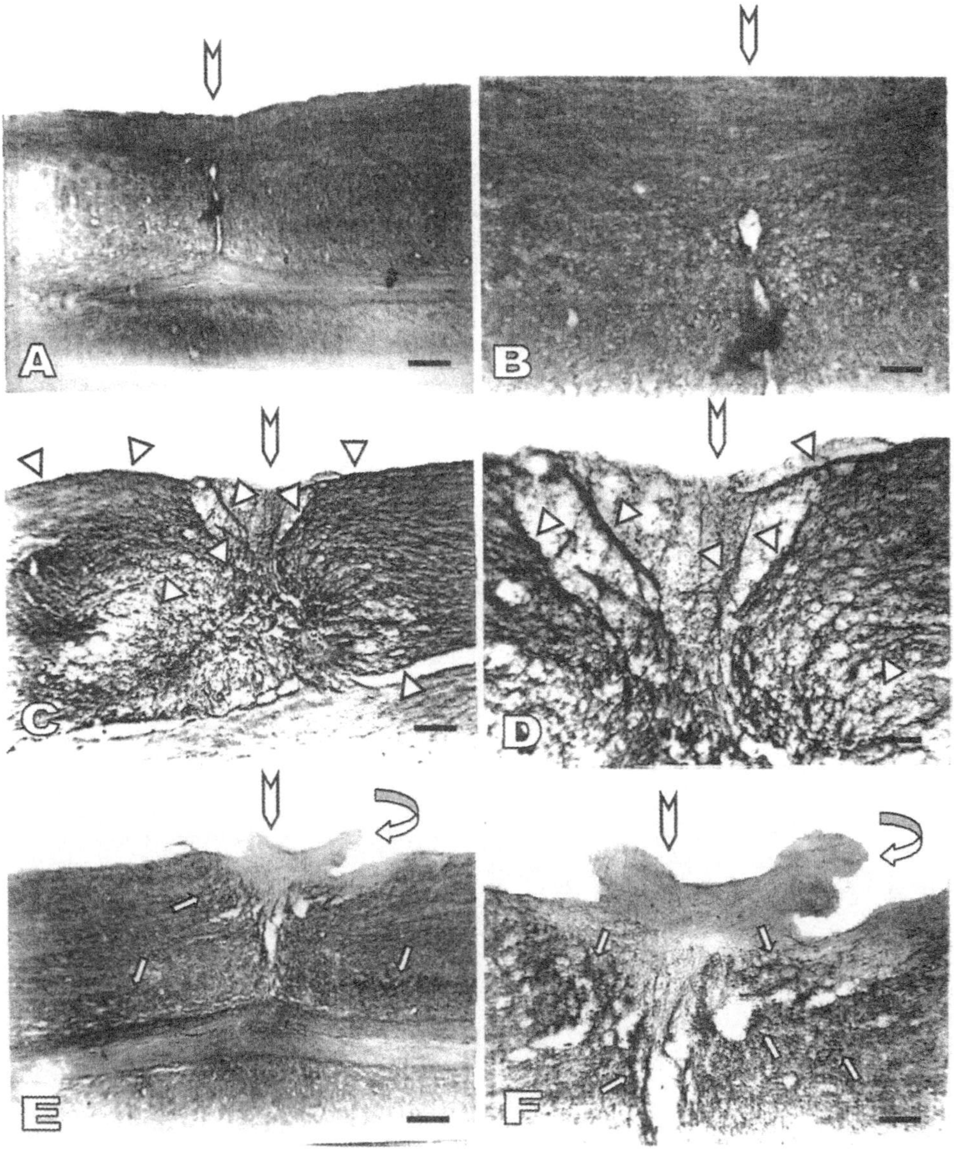

Figure 66. PRP-1-IR in the hemisected SC of non-treated (control) animals and those receiving PRP-1 for 4 weeks from the first day after hemisection. A, B: no PRP-1-IR structures are seen in the transected area (chevron); the absence of neuronal elements in injured area is noticed. C, D: PRP-1-IR nerve fibers and bundles (triangles) are evident after PRP-1 administration. E, F: in the adjacent control sections, dura matter overgrowth (curved arrow) and glial nuclei (small arrows) are seen; nerve fibers stained negatively with NGS, glial nuclei were positive. B, D, F: details of A, C, E, respectively. Bars: 100 μm (A, C, E), 50 μm (B, D, F).

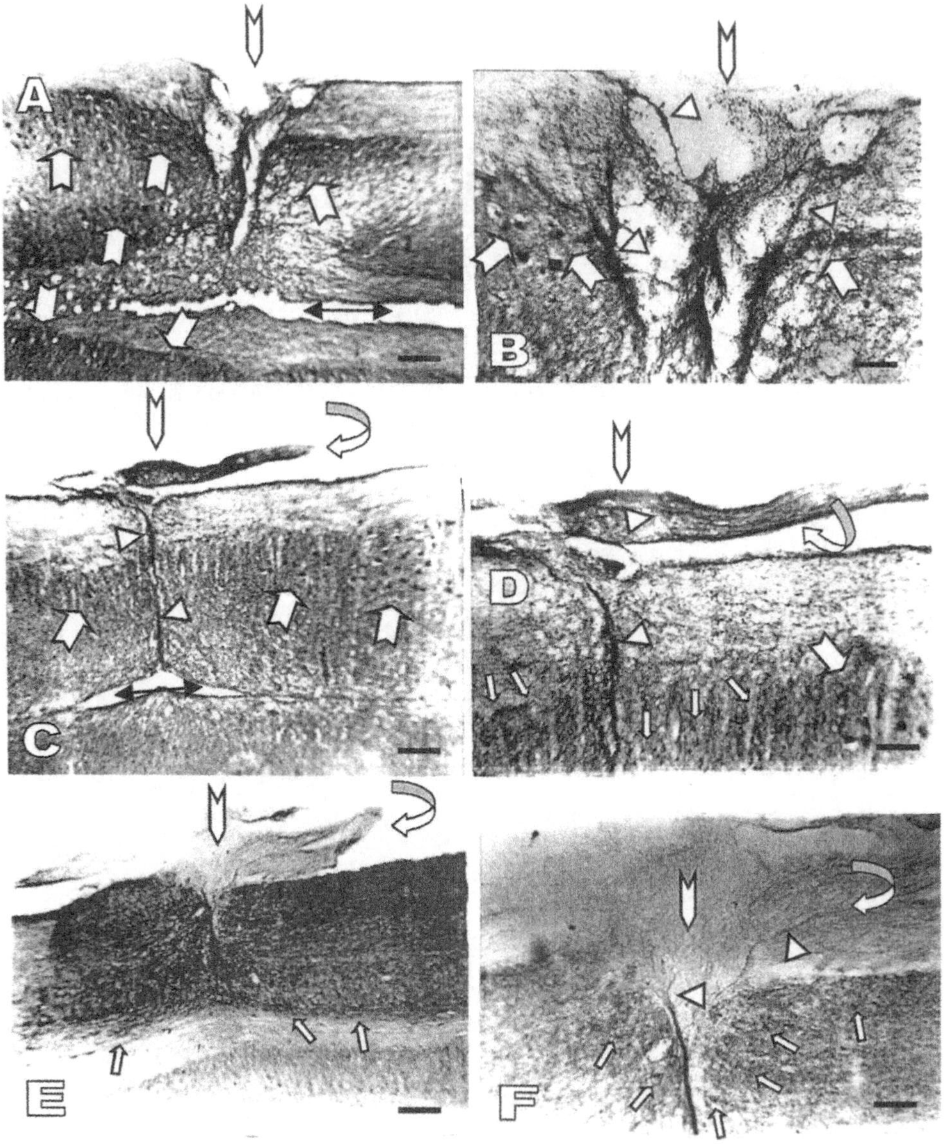

Figure 67. Administration of PRP-1 from the first day after SC hemisection for 4 weeks. The area of hemisection is indicated by chevron; in the region of dura matter overgrowth (curved arrow) the nerve fibers (triangles) show strong immunostaining for PRP-1. Under SC channel (double arrow) there is the intact side (A, C); A, B: migration of PRP-1-IR MN (large arrows) is clearly demonstrated. D: relatively high density of PRP-1-IR gliocyte nuclei (small arrows) is seen throughout the region of MN. E, F: in control sections nerve fibers stained negatively with NGS, while glial nuclei are positive. B, D, F: detail of A, C, E, respectively. Bars: 100 μm (A, C, E), 50 μm (B, D, F).

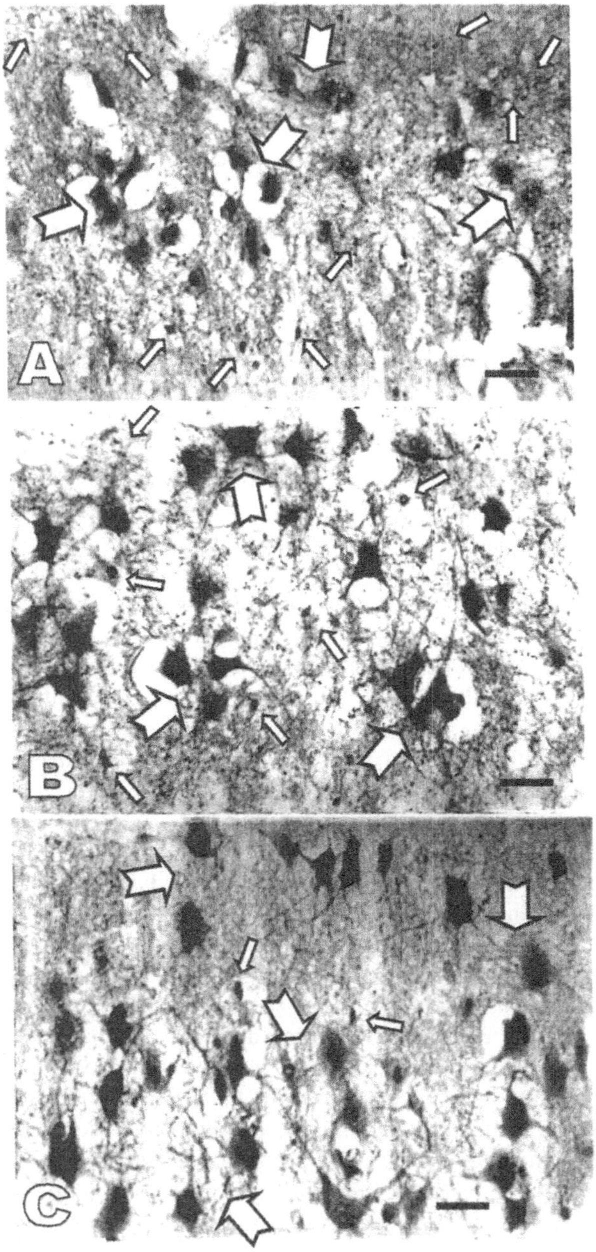

Figure 68. PRP-1-IR of SC neurons in injured area. A: degeneration of the PRP-1-immunoreactive MN of animals subjected to SC hemisection without PRP-1 treatment. In addition, PRP-1-IR gliocyte nuclei (small arrows) and fine reaction products are seen around the immunopositive degenerated MN. B, C: different stages of MN survival. Heavy staining of PRP-1-IR perikarya and processes is seen in animals after PRP-1 administration. Bars: 25 μm.

6.3. MORPHOLOGICAL AND BIOCHEMICAL CHANGES IN BRAIN MICROSTRUCTURE IN CRUSH SYNDROME AND EFFECT OF PRP-1

The crush of soft tissues (trauma of skeletal muscle caused by a long-time compression) is followed by acute hemodynamic shock, myoglobinuria, and acute renal insufficiency[251,255].

In prolonged crush syndrome of rat sciatic nerve, the disturbance of expression of mRNAs encoding proinflammatory cytokines (IFN-γ, IL-1β, IL-12 (p40 subunit), and immunosuppressant IL-10) was demonstrated by semiquantitative reverse transcription - polymerase chain reaction. Subsequently these changes promoted the development of autoimmune neuritis. Wallerian degeneration was induced along with the crush of distal stumps of sciatic nerve. After the nerve crush, both IL-1β, and IL-10 mRNAs were rapidly upregulated in the distal stump at day 1 and slowly downregulated during next 2 weeks. In the case of Wallerian degeneration, rapid up-regulation suggested a fundamental role of cytokines in the biology of the peripheral nerves[256]. Toxic products of protein nature are connected with arginine derivatives forming so-called "arginine proteins" which pass into the brain through peripheral nervous system[257].

There are numerous literature data showing that the intoxication of the organism occurs mostly during decompression when toxic metabolites are released into the blood from damaged tissues, reaching the brain due to the damage of blood brain barrier. Available clinical data show that the mortality is most frequently caused by the intoxication of the organism starting from decompression.

For example, in 4 days after crush, NO-dependent critical teleangiectasia takes place in crushed muscle as a result of iNOS activation, accompanied with intensification of the hypovolemic shock[258].

In experiments with neuronal NOS knockout of mice, the detrimental role of NO in cerebral infarction was demonstrated[259].

The general intoxication of the organism in crush syndrome is accompanied with a sharp depression of immunity. During the crush, profound neurodegenerative changes take place in the brain. This is one of the reasons for studying the effect of neuroprotector PRP-1 on morpho-biochemical changes during crush syndrome[259-261].

There was much discussion about the extent to which the brain is immune-specialized organ. Host defense mechanisms do appear to be present in the brain, since activated T-cells have the ability to cross the blood brain barrier. Early observations support the concept of blood brain barrier and indicate that the barrier extends to peripherally circulating lymphocytes. The connection of blood brain barrier with the components of the immune response has yet to be fully understood. However, CNS infection by bacteria, viruses, parasites, and the action of toxic agents, which take place also in crush syndrome pathogenesis, can produce marked CNS damage, either by direct actions or by indirect mechanisms. These agents may produce alterations in blood brain barrier permeability, edema, and elevated intracranial pressure resulting in secondary neurological deficit.

The number of publications on traumatic toxicosis in crush syndrome (CS) is rather limited, and the influence of some pharmacological preparations on the development of CS is not clearly defined. The research on CS pathogenesis has shown that major changes in CS occur during decompression. As we have previously shown, the decompression is accompanied by dramatic changes in brain energetic metabolism, glucose utilization, protein synthesis, activity of adenylic enzyme compounds, and morphological changes in brain structures[260,261].

For the experiments, 20 outbred Wistar male rats (160-200 g) were used. CS was induced by 2-hour compression of femoral muscle under a special press with a force of 100 g/kg of animal weight. The animals were divided into 5 groups: intact group, control group (2-hour of compression), and three experimental groups (2, 24, and 48 h of decompression). Each group consisted of 4 rats.

For *in vitro* study of the level of [^{14}C]-glucose utilization, brain tissue was taken out after the decapitation, perfused with 0.15 M solution of KCl, and subsequently prepared as 0.5 mm slices. The slices were weighed and placed into special containers, supplied with a central isolated funnel. Shortly, the method included the quantitation of $^{14}CO_2$ produced upon oxidation of [^{14}C]-glucose. *In vitro* incubation of the slices was conducted in 2.0 ml of Krebs bicarbonate buffer with pH 7.2, at 37°C for 60 min.

The quantity of PRP-1 used was 10^{-6} M; this quantity was found to be the most effective in preliminary experiments. 100 μl of phenylethylamine was poured into the central funnel for $^{14}CO_2$ absorption. One μCi of [^{14}C]-glucose with a specific radioactivity of 240 mCi/mM was used. The reaction was stopped by injection of 100 μl of 2N $HClO_4$ through a piece of silicone tubing passing through a hole in the container lid, and $^{14}CO_2$ was liberated. After the incubation, the funnel contents were transferred into scintillation flasks, rinsed twice with 0.2 ml of absolute ethanol, and a toluene-based liquid scintillation fluid (10 ml) was added. The method of $^{14}CO_2$ registration was described by G. Kevorkian et al. (2002, Neurochemistry, RAS & NAS RA, 19:311-317).

Creatine kinase activity was detected *in vitro*, in brain slices and blood, by spectrophotometric method with "Delta" Chemical Co. kits (Armenia).

The level of protein synthesis in subcellular particles of the brain: mitochondria (M) and endoplasmic reticulum (ER), as well as in cytosol (C), was detected *in vivo* with i/p injections of [^{14}C]-leucine, 50 μCi/100 g of animal weight (specific radioactivity 240 mCi/mM). The subcellular fractions of the brain were prepared by differential centrifugation. The purity of subcellular fractions was controlled by morphological and enzymatic methods. 20 μg of PRP-1 were injected intraperitoneally twice: immediately after the end of compression and 1 h later.

For the morphometry, tissue slices were fixed in 10% formalin and embedded in paraffin. Four-micron brain tissue slices were stained by hematoxylin and eosin accordirg to Selye. The radioactivity was measured by a liquid scintillation spectrometer SL-4221 (Roche Bioelectronique Kontron, France). Counting efficiency for ^{14}C radioactivity was 90%, δ_t - 92%. Data were expressed in cpm/mg of fresh tissue and dpm/mg of protein. Statistical analysis was conducted with Fisher-Student's method and Mann-Whitney's U-test.

Protein synthesis in brain subcellular particles and soluble fraction. It has been already shown that in crush the protein content considerably changed in liver and kidney subcellular particles and soluble fraction. Significant changes of protein synthesis were observed in brain tissue. During 2-hour compression the protein synthesis decreased in cytosol by 37.7%. In 2-hour decompression the protein synthesis decreased in cytosol by 53.7%; in 2-hour compression the synthesis decreased by 68.8% in comparison to that in the intact group (Table 19).

Under the influence of PRP-1, the protein synthesis in brain cytosol increased 5-fold in comparison with that of experimental group of animals without PRP-1, and by 54%

and 128%, as compared to intact and control groups, respectively. In crush, no reliable changes were observed in brain ER, whereas in 2-hour decompression the synthesis of membrane proteins decreased by 51.9%.

Under the action of PRP-1, 2.55-fold activation of protein synthesis in ER membranes occurred, in comparison to that with 2-hour decompression group of animals. The level of protein synthesis under the action of PRP-1 is comparable to the level of protein synthesis in intact animals.

More prominent changes were observed in membranes proteins of the brain.

During 2-hour compression the protein synthesis decreased two-fold. In 2-hour decompression it decreased up to 42.7% of that of the group of intact animals. Under the influence of PRP-1, the protein synthesis increased up to the intact group level (Table 19). In crush, a non-significant activation (by 17%) of the total protein synthesis in brain cells was registered. Total protein synthesis decreased by 56% in 2-hour decompression and increased by a factor of 3.4 under the influence of PRP-1.

Table 19. The level of protein synthesis in brain subcellular particles after 2-hour compression and 2 h decompression under the influence of proline rich peptide PRP-1.

Subcellular particles	Intact group	Control group 2 h compression	Experimental group (decompression)	
			2 h without PRP	2 h with PRP
Cytosol	10378 ± 320	6987 ± 325 p1 < 0.001	3239 ± 124 p1 < 0.001 p2 < 0.001	15962 ± 887 p1 < 0.001 p2 < 0.001
Endoplasmic reticulum	8223 ± 92	8190 ± 592 p1 < 0.5	3953 ± 318 p1 < 0.001 p2 < 0.001	10668 ± 56 p1 < 0.005 p2 < 0.001
Mitochondria	6922 ± 352	3537 ± 185 p1 < 0.001	2956 ± 153 p1 < 0.001 p2 < 0.05	7539 ± 204 p1 < 0.001 p2 < 0.001

Data are given in d.p.m./mg of protein, n=6

p1 – in comparison with intact and control (2 h compression) groups.

p2 – in comparison with control and experimental (2 h decompression) group.

Utilization of [^{14}C]-glucose in brain slices. Under the action of PRP-1, [^{14}C]-glucose utilization was stimulated by 66% in the slices of large hemispheres of intact animals. After 2-hour compression glucose oxidation was inhibited by 36% (Fig. 69). Glucose utilization rate in decompression was inhibited in 2 and 24 hours by 81% and 40%, respectively, in comparison with intact animals, and was equal to the level in intact animal after 48 hours. In 2-hour decompression 10-fold stimulation of glucose utilization was observed under the action of PRP-1. The stimulation of glucose utilization under PRP-1 action was registered in 24 and 48 hours after decompression; it was is significantly higher than in the control group (2-hour compression).

Thus, in the brain, both in normal state and in crush syndrome, a powerful stimulating effect of hypothalamic PRP-1 on glucose utilization takes place. These data provide evidence on the existence of a strong factor of regulation of aerobic and probably anaerobic glycolysis in the brain. The experiments in this direction are in progress.

Creatine kinase activity in the brain and blood. One of the main pathways of energy-rich compounds synthesis is ATP resynthesis via creatine kinase reaction. This is one of the central pathways of ATP generation in the bioenergetics.

In crush, creatine kinase activity in the blood decreased 3 times after 2-hour compression and increased non-significantly in after the 2-hour decompression. Under the influence of PRP-1, these parameters increased non-significantly (Fig. 70).

In brain tissue the following changes of creatine kinase activity were observed: 7-fold decrease after 2-hour compression and 4-fold decrease after 2-hour decompression (Fig. 70). Under the influence of PRP-1 the activity of creatine kinase in brain tissue restored almost up to the intact brain level after 2-hour compression (7-fold activation) and did not considerably elevate after 2-hour decompression.

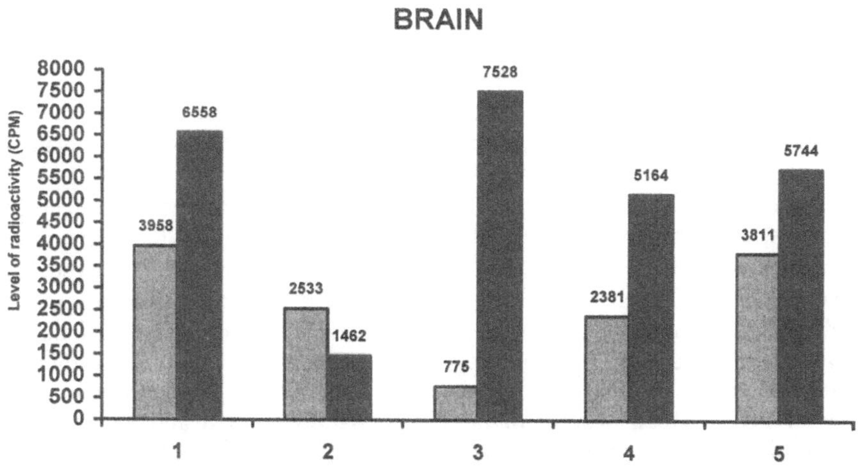

Figure 69. Glucose utilization during 2-hour compression and 2, 24, and 48h decompression with and without injection of PRP-1. 1 – Intact rats; 2 – 2-h compression; 3 – 2-h compression followed by 2-h decompression; 4 – same, 24-h decompression; 5 – same, 48-h decompression. Grey columns, without PRP-1; black columns, under influence of PRP-1 (10^{-6} M).

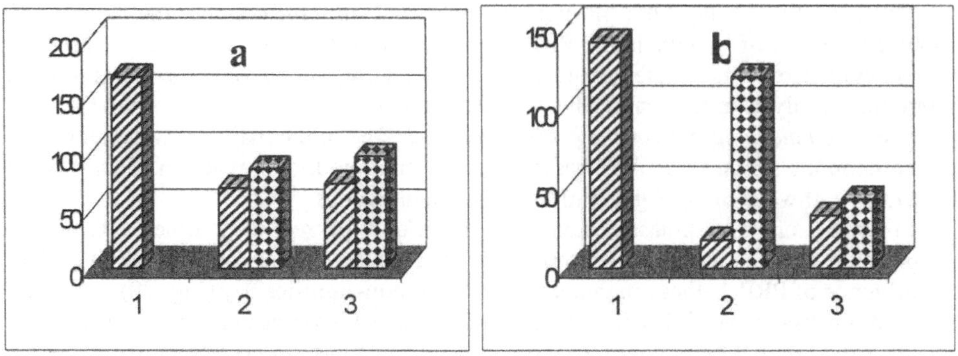

Figure 70. a - Creatine kinase activity in rat brain in crush syndrome with and without PRP-1; b - creatine kinase activity in rat blood in crush syndrome with and without PRP-1. 1 - intact animals; 2 - control animals: 2-h compression (dashed) and 2-h compression + PRP-1 (diamonds); 3 - experimental animals: 2-h compression + 2-h decompression (dashed) and 2-h compression + 2-h decompression + PRP-1 (diamonds).

Morphometric investigation of brain structures. After 2-hour decompression, the brain capillaries are collapsed and free of blood cells. Single plasmatic vessels are seen. Separate venules are dilated and full of blood. Perivascular edema is totally observed. In CG1 and CG2 fields of all cortical layers, central, peripheral and total chromatolysis is seen. Sharply expressed chromatolysis of neurons of II, III, and IV cortical layers occurs in Par 1 and Par 2 fields. Groups of neurons with features of vacuolization, chromatolysis, and phagocytosis are present in V layer. Besides, there are groups of hyperchrome and pyknotic neurons. Subcortical neurons are mainly preserved. Only in separate neurons the hyperchromatosis, pyknosis, neurophagia, and chromatolysis are observed. Longitudinal bundles of Cpu are considerably edematous; local gliosis and total glial edema appear in Cpu. The vessels of vascular plexi of lateral ventricles are with collapsed empty lumens (Fig. 71). Locally, after 2-hour decompression under the influence of PRP-1 small arteries of *pia mater* are constricted and empty. Moderate amount of blood cells is present in veins. Brain capillaries are functional. Separate plasmatic vessels are observed. As a rule, chromatophilic substance in neurons of every cortical layer is thin-dispersed and preserved in CG1 and CG2 fields. Only in separate neurons a slight central chromatolysis is observed. In Par 1 and Par 2 fields, the integrity of neurons throughout all layers is also maintined. Separate hyperchrome neurons are detected in V layer. Locally, in II, III and IV layers, slight peripheral chromatolysis is visible, whereas in separate neurons total chromatolysis is observed. Phagocytosis of single neurons was detected in local areas of IV and V layers. Subcortical neurons are well preserved. The edema of longitudinal bundles and Cpu gliosis are not detected. Vessels of vascular plexi of III and lateral ventricles contain blood cells in moderate amount (Fig. 72).

The development of CS is accompanied with a sharp decrease in energy substrates. The study of CS in decompression period deserves special attention. Since this period of disease, the distribution of toxic products takes place in the whole organism along with the development of irreversible damage of metabolism. This fact served for designing next series of our experiments in which we compared the animals in decompression with those undergone compression only. In decompression period anaerobic glycolysis becomes the main energy source. The brain generates ATP in large quantities to maintain the membrane potential essential for the transmission of nerve impulses. Under normal conditions the brain uses glucose only to meet its prodigious energy requirement, which amounts to about 60% of the glucose utilization in humans at rest. The brain needs approximately 120 g of glucose per day, having no significant glycogen or other fuel reserves. The supply of both oxygen and glucose cannot be interrupted even for a short time. According to our data, this period is characterized also by a sharp decrease in uptake of glutamic acid and synthesis of proteins due to low level of ATP[251]. The sharpest suppression of the glucose utilization is observed in the brain after 2-hour compression. Thus, the initial period after compression is the most difficult for the organism because of intoxication, and the protective mechanisms are not capable to bring ATP generation to the initial level.

The phenomenon of crush-induced intoxication was also described after the earthquake in Armenia in 1988, when the highest percent of mortality was observed in first hours after the release from blockages. Later, the organism gets adapted, and the endogenic protective mechanisms begin to restore. Thus, more than two-fold increase in glucose content was registered in crush. In this condition PRP-1 can positively influence on brain survival by delivering energetic substrates via blood to cerebral tissue. Thus, the hypothalamic peptide PRP-1 has the most positive and favorable effect on glucose utilization by cerebral tissue, increasing its level in the blood; this effect compensates a 2.4-fold decrease in blood insulin which takes place in CS[260].

Despite of the fact that the intoxication of the organism starts with decompression, bioenergetic mechanisms in the brain undergo sharp changes from the start of compression. So, creatine kinase activity lowers significantly durind compression, whereas the oxidation of glucose varies not so much. The period of compression is characterized by the beginning of hypoxia and cerebral ischemia. Both aerobic and anaerobic mechanisms of energy production are inhibited in this period; high quantity of glucose and high creatine kinase activity are detected in the blood. In 2-hour decompression their level in the blood does not vary, and the mechanisms of energy production in brain tissue are inhibited stronger. The effect of PRP-1 is realized through the regulation of both aerobic and anaerobic pathways of energy generation. They are activated at different stages of crush syndrome development.

In decompression, anaerobic glycolysis becomes the only way of energy generation. We demonstrated that in hypoxia PRP-1 administration provides brain tissue with energy supply to due to the enhancement of anaerobic glycolysis.

The disturbance of protein synthesis is also due to energetic deficit. As it follows from our experimental data, the synthesis of cytosolic and mitochondrial membrane proteins is strongly inhibited during compression. The inhibition of protein synthesis increases in decompression. In this period, the inhibition of protein synthesis is observed in all studied organelles of brain tissue. PRP-1 restores the system of energy production for protein synthesis in all cellular fractions. Therefore, the depression of protein synthesis depends on the absence of high-energy compounds. The reinforced degradation of proteins observed in crush syndrome promotes an increase in free amino acid pool and peptides due to the natural degradation of proteins, as well as to the low availability of energy for their resynthesis.

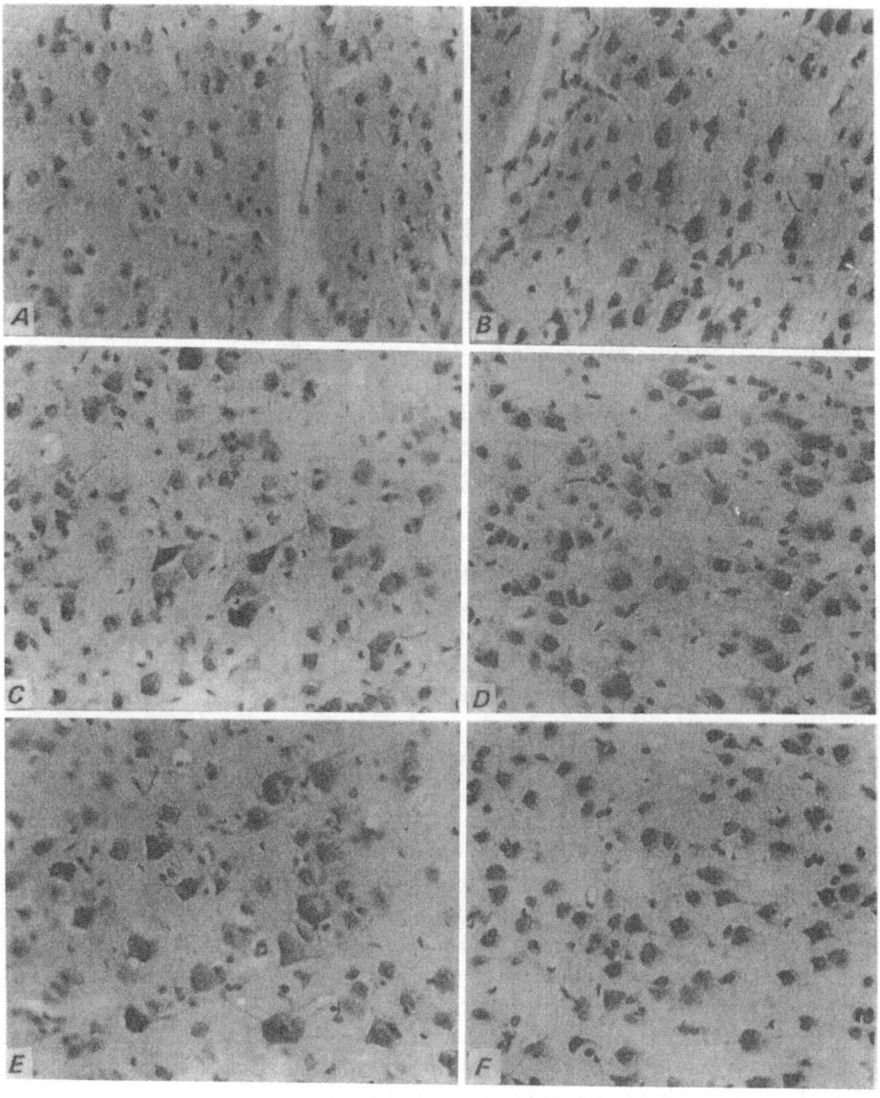

Figure 71. Rat brain after 2-hour compression and subsequent 2-hour decompression. A – central peripheral total chromatolysis of neurons in CG1 and CG2 fields. The capillaries are collapsed. B – CG1 and CG2 fields under action of PRP-1. Chromaphilic substance is mainly preserved, in separate neurons moderate chromatolysis is noticed. Capillaries and venules keep their lumen. C – neuronal pyknosis and neurophagia in V cortical layer of Par.1 field. D – after administration of PRP-1 chromaphilic substance is preserved in the neurons of the same field and is thin-disperse. Single hyperchrome neurons are seen. E – in V cortical layer of Par.1 field the group of neurons with cytoplasm vacuolization and chromatolysis is observed. F – neurons are well maintained under PRP-1 action. Staining by Nissle. Magnification x 400.

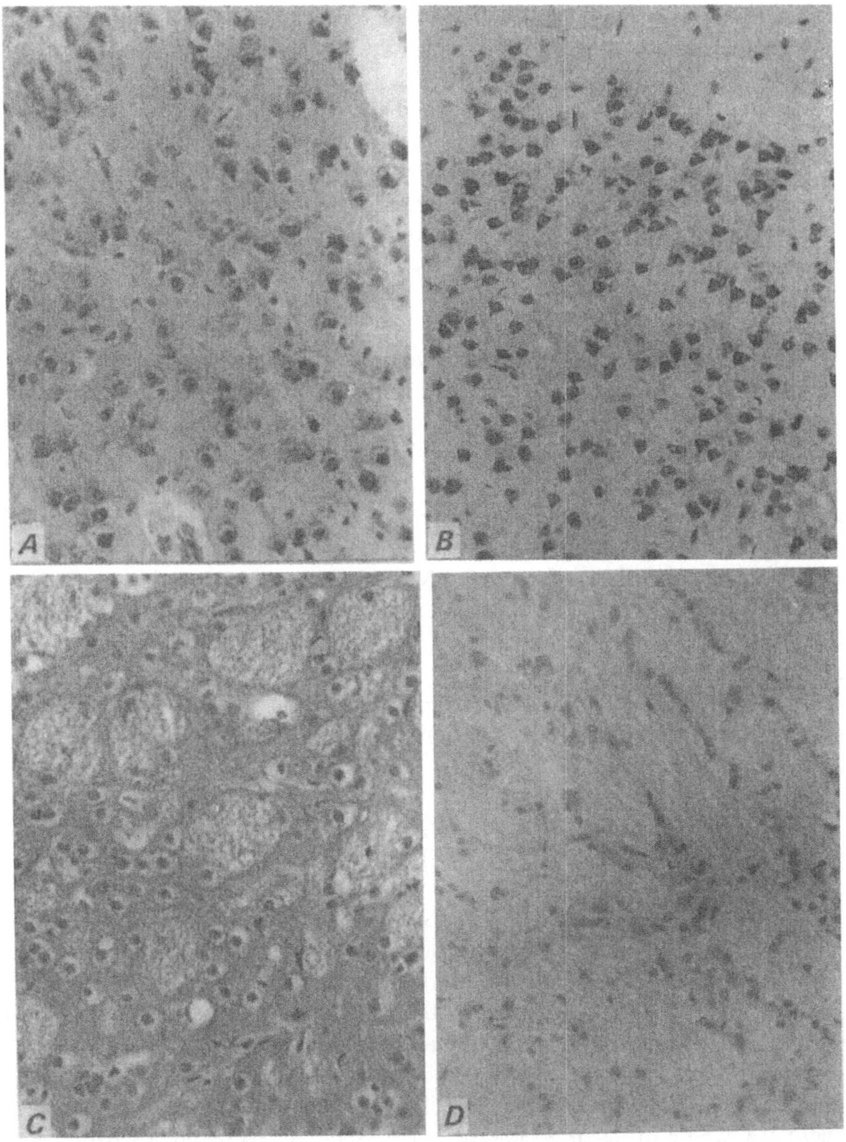

Figure 72. Rat brain after 2-hour compression and subsequent 2-hour decompression (effect of PRP-1). A – extended, predominantly total chromatolysis of neurons of II and III cortical layer of Par.1 field. B – local chromatolysis in relevant zone under PRP-1 action, chromaphilic substance is partly preserved and thin-disperse. C – expressed edema of the longitudinal bundles, local gliosis and glial edema in Cpu is observed. D – expressed edema of longitudinal bundles Cpu, but glial edema can not be detected. Staining by Nissle. Magnification x 400.

The positive effect of PRP-1 depends, in all probability, on the activation of immune system and adaptation mechanisms, including the mobilization of endogenic protective mechanisms of the organism itself.

The electron microscopy studies carried out by us indicate that 2-hour compression results in increasing the sizes of cell organelles (in comparison with those of intact animals).

The shape of mitochondria is round, and matrix is lightened. In most of M large spots of focal lysis are observed. The outer mitochondrial membrane is maintained, though it sometimes looks destroyed. The crysts loose typical parallel arrangement and adopt a honeycomb-like configuration. Certain M are transparent, and besides normal crysts, honeycombs-like structures are observed. A part of M is irreversibly damaged.

An increase in distances between the crysts and respiratory ensembles located on the crysts occurs, which is accompanied by uncoupling of oxidative phosphorylation reactions (Figs. 73, 74, 75).

In the period of 2-hour decompression followed by 2-hour compression the M swell. Their sizes increase strongly, and they are round in shape. Outer membrane of some mitochondria is damaged. Washing out mitochondrial matrix takes place. Often mitochondrial configuration is of honeycomb type, but they are more swollen, and the decay of their structure occurs.

Introduction of PRP-1 on the background of 2-hour decompression, which followed 2-hour compression, is accompanied by transforming of brain cell morphology from damaged to intact type. Thus, the size of M decreases, the matrix is maintained in the majority of the organelles, and its electron density is higher, but in some cases the matrix is still washed out. The outer membrane is maintained, but in rare cases it is still destroyed.

The restoration of cryst parallel arrangement, disappearance of swollen M, and normalizing the disposition of respiratory ensembles on the crysts is characteristic of this series of experiments (in press).

6.4. PRP-1 PROTECTS BRAIN NEURONS AGAINST ALUMINUM (Al) NEUROTOXICOSIS (ELECTRON MICROSCOPY STUDIES).

In 1980 H. Wishnevski and co-workers[262] studied brain nerve cells of animals with modeled Al neurotoxicosis by electron microscopy and demonstrated the presence of double helical neurofilaments (not observed for a long time before), which were characteristic of Alzheimer's disease (AD). In further studies these structures were discovered in the cytoplasm of cerebrospinal ganglion neurons of old rats. Since 1983 till now electron microscopy studies of experimental models of $AlCl_3$ neurotoxicosis have been carried out by Prof. A. P. Avtsin and his co-workers[263].

There are twelve forms of Al microelementosis[264,265]. Al competes with calcium and magnesium and displaces them from cell membranes. Damaging effect of Al ions on the organism is widely investigated in laboratories and clinics in connection with its role in amyloid formation and potential role as an agent in AD etiology[266]. The association of transferrin (TF) isoforms C1 and C2 correlates with a loss of its ability to bind and transfer ions of Al. The latter, in its turn, enhances free-radical oxidative processes with participation of oxygen active forms (OAF) in different organs and tissues of the organism, including the brain[267,268]. The oxidative damage is one of the mechanisms of neurodegeneration. Ions of iron and Al are also involved in this process, participating in the cleavage of hydrogen peroxides, products of enzymatic dismutation of superoxide radicals (OH) in brain tissue. Experimental data exist which indicate the role of Al in the pathogenesis of different human diseases, including AD[267,269,270].

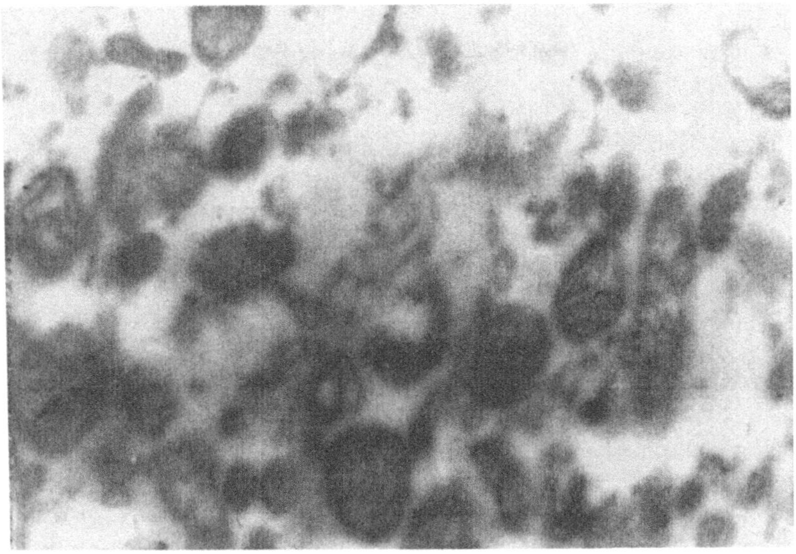

Figure 73. Structure of brain mitochondria in intact animals. Magnification x 24 000.

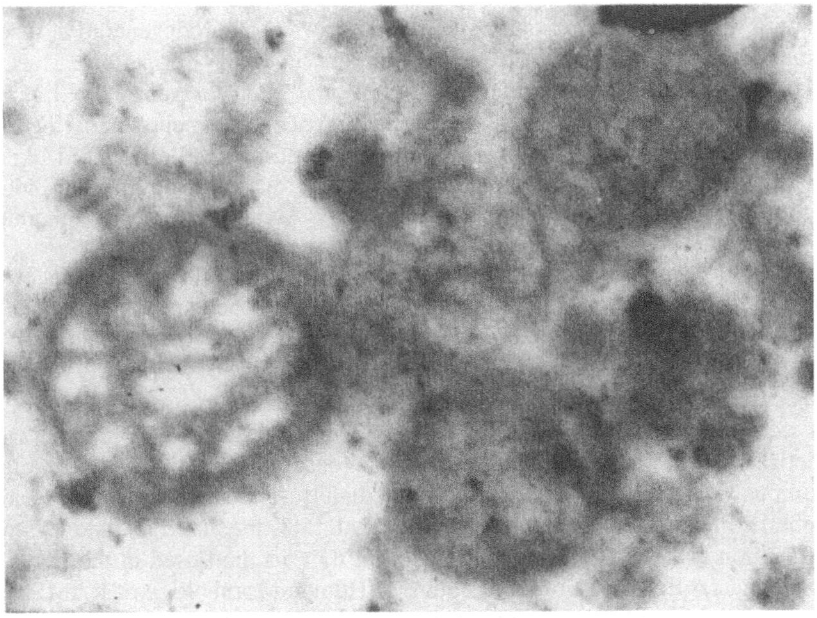

Figure 74. Structure of brain mitochondria after 2-h compression and 2-h decompression. Magnification x 30,000.

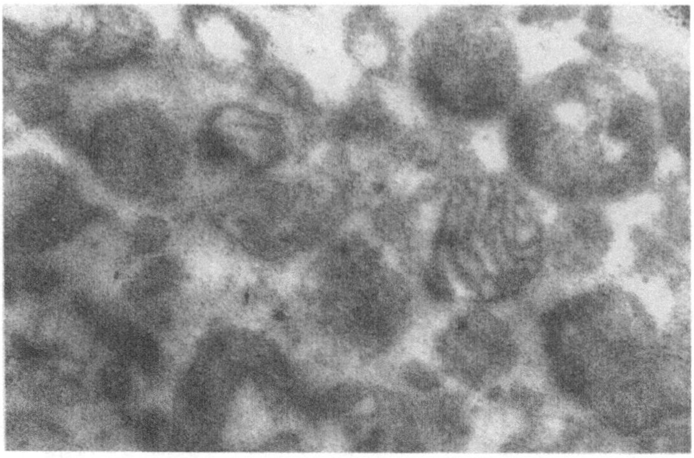

Figure 75. Structure of brain mitochondria after 2-h compression and 2-h decompression under influence of PRP-1. Magnification x 24, 000.

We demonstrated that $AlCl_3$ produces degenerative changes in the ultrastructure of L_{929} fibroblasts and Hasser's bundle neuroma cells *in vitro*[271] (see Chapter VIII). It is worth noting that one of important pathogenic factors of neurodegenerative disorders is the disturbance of mechanisms of neurosecretory cytokine formation. Thus, certain cytokines produced in the brain *in situ* (in NPV and NSO) are considered as inducing factors for the whole cascade of immunopathological and neuroendocrine disorders resulting in the accumulation of atypical proteins in neuronal cells, amyloid-like masses. In view of above mentioned facts, studying the effect of PRP-1 on morpho-biochemical changes in the brain and other organs induced by $AlCl_3$ was both of theoretical and practical interest[272].

White adult rats of both sexes (150-180 g) were divided into 4 groups by 10 animals each. In experimental group-1 (EG-1), animals received subcutaneous injections of 2 ml of 10% $AlCl_3$ solution, done on the first and third day; in EG-2, the animals were injected with 25 μg of PRP-1 one hour before receiving the first dose of Al; in EG-3, the animals were injected with 25 μg of PRP-1 6 days after the Al administration. Control animals (EG-4) were injected 3 times with 0.2 ml of 9% NaCl instead of Al and PRP-1. After anaesthetization with ether, the animals were decapitated. Blood, liver, brain, and hippocampus were taken out and treated immediately. Tissue slices were prepared from the cerebral cortex of hippocampus, large and small intestine, liver, and kidneys within of 3 min. Fixation and embedding were carried out by a method used in the Laboratory of Cell Experimental Pathology of the Institute of Human Morphology of RAMS. Ultrathin slices were studied by electron microscope JEM-100 CX at 80 kV accelerated voltage.

One hour after intraperitoneal injection of 0.2 ml of 10% $AlCl_3$, edematic neurons were observed in experimental animals' hippocampus by electron microscopy (Fig. 76a, b, c)[272]. Moreover, neurons of light and dark texture of cytoplasm were observed at the same time (Fig. 77). After 3 daily injections of 10% $AlCl_3$ solution, some neurons with diffuse distribution of discrete Al particles in the cytoplasm were observed in the hippocampus. In some cells small lumps of granular electron-dense particles were observed (Fig. 78); in some cases, besides the small lumps, diffuse penetration of small granular particles through nuclear membrane was discovered (Fig. 79). Noteworthy, in some myelin nerve fibers accumulation of Al took place (Fig. 80). After preliminary administration of PRP-1 by A. Galoyan's scheme, neither neurons with cytoplasm and nuclei edema, nor neurons containing electron-dense granules and large inclusions in the cytoplasm or nucleus were revealed in the hippocampus (Fig. 81).

We also studied the epithelial cells of small intestine. Rapid penetration of Al into the cytoplasm of the epitheliocytes in micropinocytic vesicles was established. By the third day they appeared between inner and outer nuclear membranes (Fig. 82). In the same experiments Al granules were discovered in kidneys between epithelial cells in intraepithelial holes in Henle's loop (Fig. 83). Similar picture was observed with glandulary epithelocytes of stomach (Fig. 84) and liver cells (Fig. 85). After preliminary administration of PRP-1 we revealed no Al granules in all cell types studied: hippocampus neurons, epithelia of liver, kidney, small intestine, and gastric gland.

Thus, the obtained data provide evidence that Al is capable of penetrating the cytoplasm and cell nuclei, where it interacts with cellular DNA.

Earlier we studied the effect of $AlCl_3$ on nerve cells in the culture of RGGN cells[272] established in our laboratory. Later on, the chromosomal apparatus and chromatids were studied with G- and C-techniques. Data obtained in those studies convinced us that the genetic apparatus of nerve cells did not undergo any changes in the presence of $AlCl_3$. So, Al does not possess mutagenic properties but disarranges the formation of mitotic spindle and cytoplasmic complex of microtubules. Al ions penetrate the cells (nerve and epithelial) through ion channels of the plasmalemma. This phenomenon is called *cytotesaurismos*, and we call the penetration of $AlCl_3$ via nuclear pores into nucleoplasm *nucleotesaurismos*[272].

Thus, we succeeded to visualize the manifestations of Al toxicosis in rat hippocampal neurons, epitheliocytes of liver, kidneys, small intestine, and stomach, which are accompanied, probably, with the disturbance of oxidative metabolism of the whole organism. Noteworthy, our data provide an evidence that PRP-1 exerts neuroprotectory and epithelioprotectory effects in Al toxicosis.

During past years there is evidence in the literature that aluminum can be accumulated in the brain neurons and induce formation of amyloid proteins causing Alzheimer – like disease. Data obtained by us proves that PRP-1 can prevent aluminum neurotoxicity by eliminating of aluminum from its place of accumulation in the nervous system. Thus, hopefully PRP-1 can be used as remedy for treatment of Alzheimer disease. In favor of this we have other series of data testifying untineurodegenerative properties of PRP-1 during snake venom intoxication, under crush syndrome, hemisection of spinal cord etc. We had carried out the effect of PRP-1 on the morpho – functional properties of brain neuros under Beta-amyloid fragment 25-35. We are continuing our experiment in that direction.

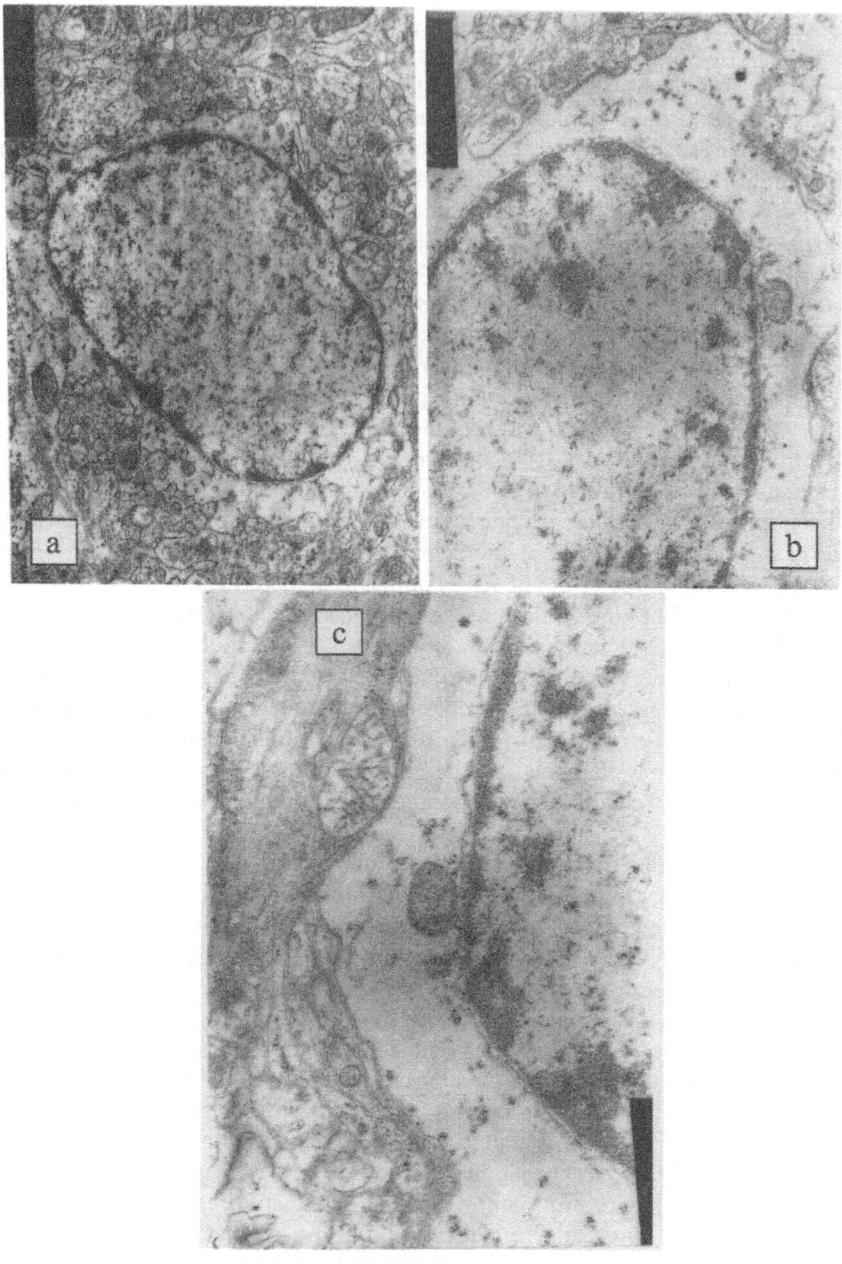

Figure 76. (a) Rat hippocampus. Intact neuron. Magnification x 10 000. (b) A neuron 1 h after injection of 10% solution of $AlCl_3$. Cell fragment with edematic cytoplasm and nucleus. Magnification x 14 000. (c) The same preparation. An edematic neuron 1 h after injection of $AlCl_3$ solution. Magnification x 20,000.

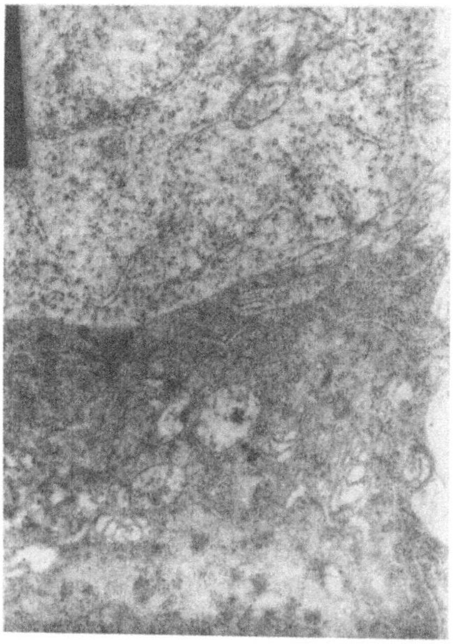

Figure 77. Rat hippocampus 1 h after injection of $AlCl_3$ solution. A detail of adjoining neurons with light and dark texture of cytoplasm. N – nuclei, C – cytoplasm. The arrow shows cell boarder. Magnification x 14,000.

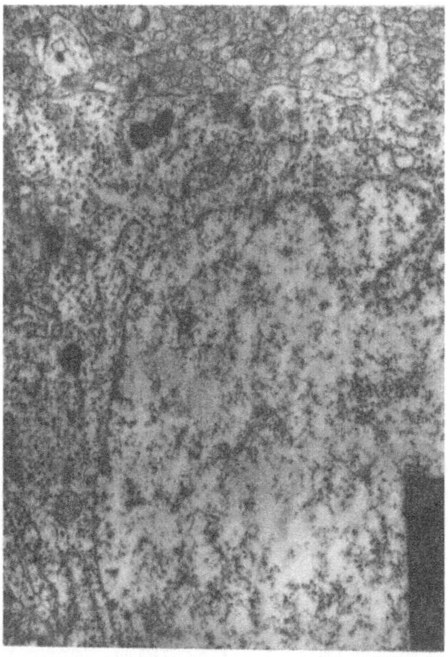

Figure 78. Rat hippocampus. A neuron after three daily injections of $AlCl_3$ solution. Electron dense particles are seen in the cytoplasm. Magnification x 10,000.

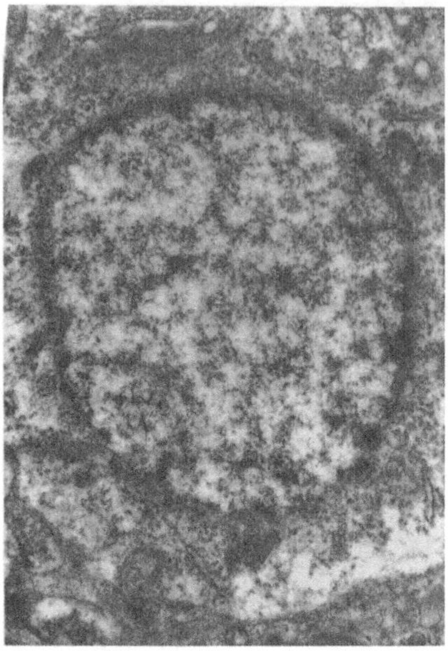

Figure 79. Rat hippocampus after three injections of $AlCl_3$ solution. Diffuse penetration of aluminum through nuclear membrane is seen. Magnification x 10,000.

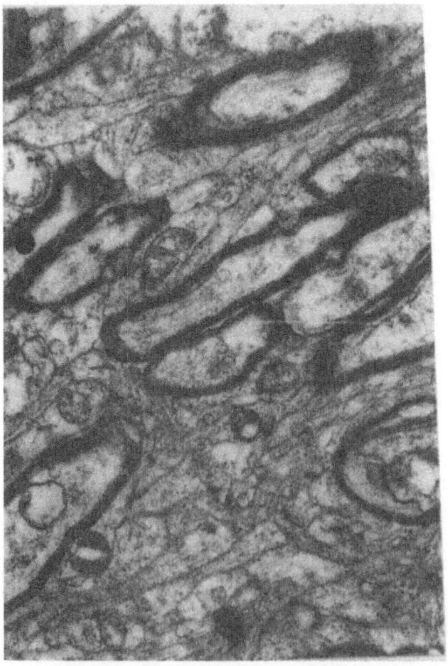

Figure 80. Rat hippocampus after three injections of $AlCl_3$ solution. Aluminum located in myelin membranes of nerve fibers. Magnification x 10,000.

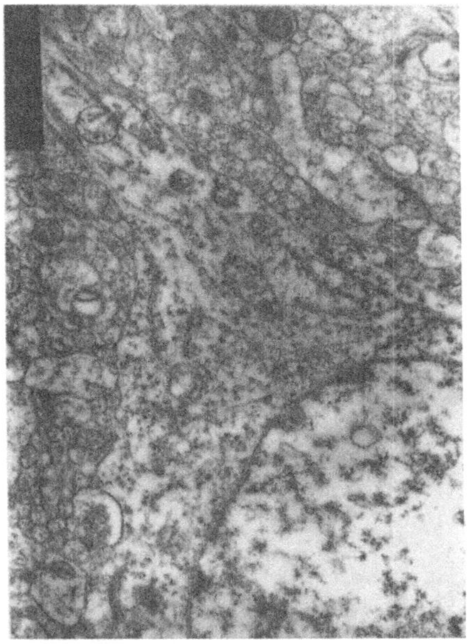

Figure 81. Rat hippocampus. A neuron after preliminary administration of PRP-1 and three injections of $AlCl_3$ solution. Magnification x 14, 000.

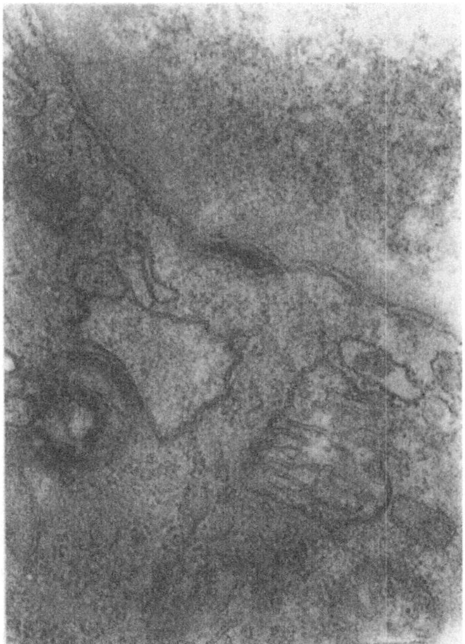

Figure 82. Epithelial cell of rat small intestine after three injections of $AlCl_3$ solution. Aluminum granules between internal and external membranes are seen. Magnification x 20,000.

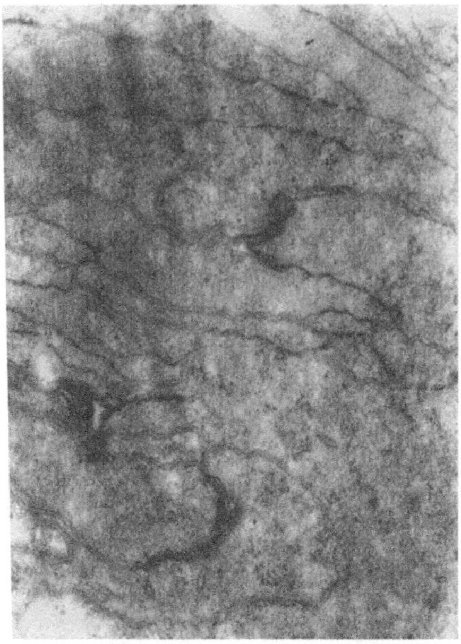

Figure 83. Rat kidney 72 h after three injections of $AlCl_3$ solution. Aluminum granules are localized in intraepithelial holes of Henle's loop. Magnification x 20,000.

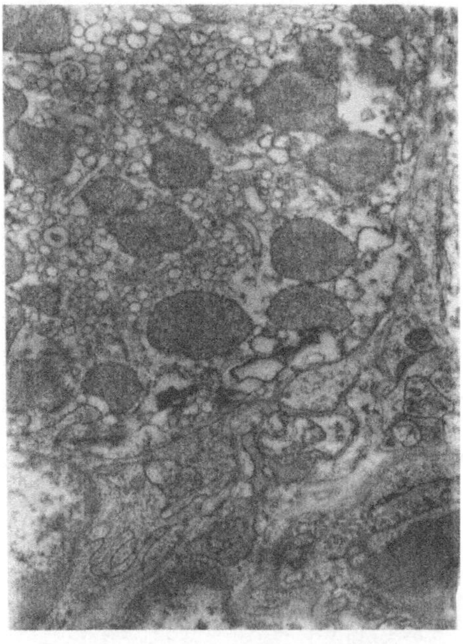

Figure 84. Rat stomach after three injections of $AlCl_3$ solution. Aluminum granules in the cells of glandular epithelium. Magnification x 13,000.

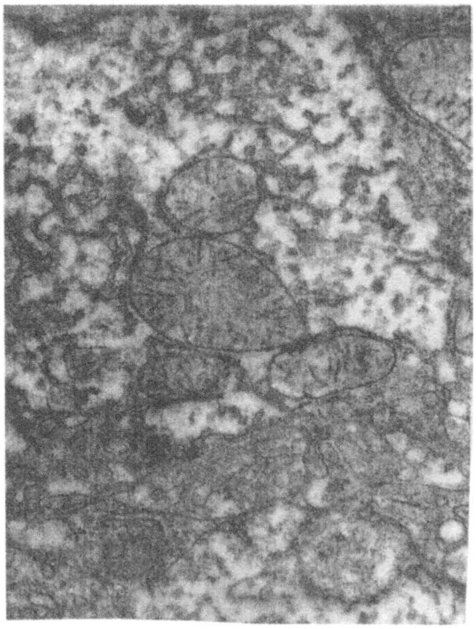

Figure 85. Rat liver after three injections of $AlCl_3$ solution. Aluminum in the cytoplasm of liver cells is seen. Magnification x 14,000.

Thus, PRP-1 protects the cells against Al neurotoxicity and abolishes sharply accumulation of this element in brain microstructures and visceral organs (liver, kidneys, stomach, and intestine). Data obtained show that PRP-1 may be recommended for treatment and prophylaxis against the aluminosis. The protective effect of PRP-1 against the Al neurotoxicity can be explained by its anti-oxidative, anti-free-radical properties.

The organism is protected against a possible pathogenic influence of Al by a number of physiological mechanisms, the most important among which is a regular urine excretion. According Avtsin[264], the following forms of aluminosis should be distinguished: 1) Simple Al accumulation in CNS, which occurs in persons over 56 years. 2) Al accumulation in AD in severe forms of presenile and senile dementia. 3) Dialysis Al encephalopathy. Many cases of dialysis dementia may result from Al accumulation in the brain. 4) Non-dialysis infantile encephalopathy. 5) Al encephalopathy in total parenteral nutrition; 6) Jatrogenic dialysis Al osteodistrophy; 7) Jatrogenic peritoneal aluminosis, etc. Presently our experiments on the subject are in progress. Thus, PRP-1 is a potent drug for treatment of Al neurotoxicity acting against accumulation of Al in different organs.

7

EFFECT OF PRP-1 ON TUMOR CELLS

For more than 30 years, in the Laboratory of Neurohormone Biochemistry we systematically studied neurospecific proteins and organotropic hormones of the hypothalamo-neurohypophysial system. In the 1970s I conceived the idea of organotropic influence of liberins (releasing hormones). Moreover, we published a series of papers on the effects of liberins on the metabolism of different organs. We studied effects of LH-RH and analogs (kindly provided by Prof. A. Schally during my visit to his laboratory) on the kallikrein-kinin system of rat blood, effects of TRH and somatostatin on oxidative phosphorylation in the mitochondria of heart, kidney, and liver of rats and rabbits after hypophysectomy[273,274], and the kallikrein-kinin system of blood after hypophysectomy[275,276], etc. Thus, according our data, the releasing hormones possess organotropic activity and metabolic activity at concentrations of 10^{-8} - 10^{-9} M. Concerning organotropic activity of somatostatin, R. Guillemin wrote: "It remains to be shown whether somatostatin is a normal physiological regulator of the secretion of GH, glucagon or insulin in mammalians, particularly in man. Regardless of the outcome of those studies still to come, availability of somatostatin opens already an entirely new conceptual approach to the understanding of secretion regulation (in normal and pathological states) of GH, insulin, and glucagon. Unsuspected extra-hypophysiotropic effects of somatostatin, a peptide of brain origin influencing peripherally the secretion of glucagon and insulin, constitute probably the most far-reaching contribution of these studies. Such an observation is in agreement with proposals made several years ago by A. A. Galoyan that hypothalamic peptides, other than vasopressin and oxytocin, may have peripheral organotropic activities" [277].

Organotropic activity of liberins and cardioactive neurosecretory hormones of hypothalamus was described by us in great detail[7,31]. Our investigations showed that the PRPs also possess organotropic activity both towards immunocompetent cells of blood and neurons of brain and SC, in normal and pathological states. For the first time we have shown a dose-dependent effect of PRP-1 on tumor cells.

Prof. A. Schally and co-workers studied the influence of analogs of releasing and inhibitory factors of hypothalamus on tumors of different organs. Studies on the role of tumoral LH-RH, growth hormone-releasing hormone (GH-RH), and bombesin/GRP, and their receptors in the proliferation of various tumors were summarized by Schally et al.[5]: "Treatment of many cancers remains a major challenge, but new therapeutic modalities are being developed based on antagonists of GH-RH and biochemistry, which inhibit growth factors or their receptors". Prof. Schally reviewed the involvement of hypothalamic hormones in the processes of malignant growth and especially the use of synthetic analogs of hypothalamic peptides for treatment of various cancers. Taking into consideration that PRP-1 produced by NPV and NSO cells is released into the general circulation and exerts its effect on the activity of immunocompetent and neuronal cells, in normal and pathological conditions, as well as on the proliferation of Jurkat cells *in vitro*, we studied the *in vitro* effect of PRP-1 on different tumor cells. Data on the effect of PRP-1 on cultured tumor cells are presented below.

7.1. EFFECT OF PRP-1 ON THE MORPHOLOGY AND MITOTIC ACTIVITY OF NEURINOMA OF RAT NODULI GASSER (NRNG) CELLS IN CULTURE (ELECTRON MICROSCOPY STUDIES)

It was of a certain interest to study PRP-1 effect on nervous tumor cells *in vitro*. Our aim was to investigate PRP-1 effect on morphology, ultrastructure, and mitotic activity of cells in culture. A unique neurinoma cell line *NRNG* obtained and supported in the Laboratory of Experimental Neurooncology of the Research Institute of Human Morphology of Russian Academy of Medical Sciences was chosen as a model for our investigations. Lately the cells were used as an adequate model for studying and diagnostics of prion disease[278].

The cell culture was grown in RPMI medium (Flow) containing 15% FCS and gentamycin (100 μg/ml). In 24 h the cell morphology and cell cycle were studied in control and in the presence of PRP-1 at concentrations of 5 x 10^{-3} M and 5 x 10^{-5} M. Glasses with cells were fixed in the Carnois solution. Staining was made by hematoxylin and eosin. Mitotic index was calculated as described elsewhere[279]. The neurinoma cells were removed using EDTA solution. Then they were then suspended in culture medium and centrifuged at 1500 rpm for 3 min. The sedimented cells were fixed by 2.5% glutaraldehyde in 0.1 M cacodylate buffer, pH 7.4, for 30 min. After that the cells were washed in 0.1 M cacodylate buffer and additionally fixed by 2% solution of OsO_4 in the same buffer, during one hour. Then the cells were washed in the same buffer and dehydrated in acetone solutions with ascending concentrations. The cells were embedded in the mixture of eponym with araldite. Ultrathin slices were made using LKB ultramicrotome (Sweden), then were contrasted by preparations of lead by Reynolds and studied by electron microscope JEM-100 B.

Light microscopy of logarithmically growing *NRNG* cells showed that the mitotic index increased 1.6 times in the presence of the neuropeptide. 1.3-fold increase in gigantic cell number was revealed in the experiment compared to control. However, the index of pathological mitoses did not elevate, but on the contrary, it was 1.6 times lower compared to control. The predominance of any definite form of pathological division was not observed. Cytomorphological picture of the culture in the presence of PRP-1 did not differ from that of intact culture (Fig. 86A). Thus, the analysis of mitotic activity of *NRNG* cells in the presence of neuropeptide PRP-1 allows us to conclude that some stimulation of the proliferative activity by this agent occurred on the background of a decreasing number of pathological mitoses and insignificant increase in the quantity of gigantic cells (Fig. 86B).

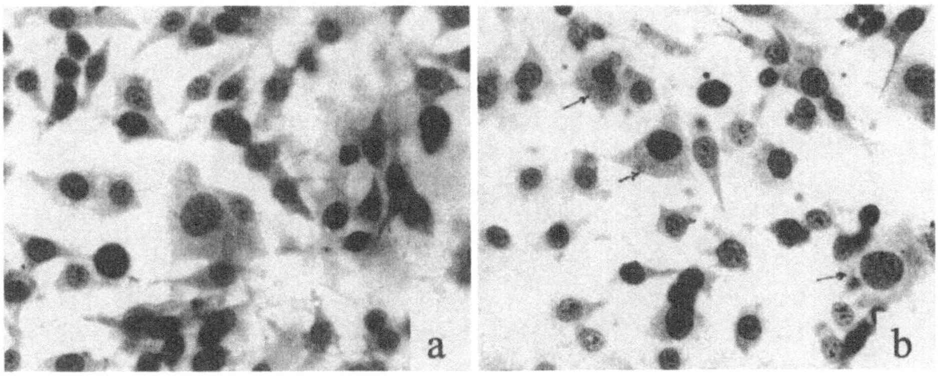

Figure 86. (a) - cyto-morphological pattern of NRGG-1 cells without PRP-1; (b) – same in the presence of PRP-1.

Electron microscopy of *NRNG* cells in the presence of PRP-1 (5 x 10^{-3} M, 24 h) revealed the cells with porous structure of cytoplasm resulted from widening of cavities of the smooth endoplasmic reticulum (Fig. 87). These changes were not observed in control (Fig. 88). In the presence of PRP-1 (5 x 10^{-5} M), fragments of destructed electron dense nuclei surrounded by membranes were observed in ultrathin slices, an indication on primary stages of tumor cell apoptosis (Fig. 89). Thus, under *in vitro* conditions PRP-1 inhibits the growth of NRNG cells at concentrations of 5 x 10^{-3} – 5 x 10^{-5} M. Porous structure of the cytoplasm resulting from extended cavities of the smooth endoplasmic reticulum was observed by electron microscopy. Fragments of destructed electron-dense nuclei surround by membranes indicated the development of primary stages of apoptosis in these tumor cells. In the case of mature cells these changes are a substantial widening of non-granular cytoplasmic reticulum, fragmentation of the nuclei and cytoplasm, and destruction of internal membranes of swollen mitochondria. As for immature cells, they are subjected to other changes in the ultrastructure: karyorrhexis and karyolysis is developed in pyknotic nuclei, and microplasmatosis is strongly expressed.

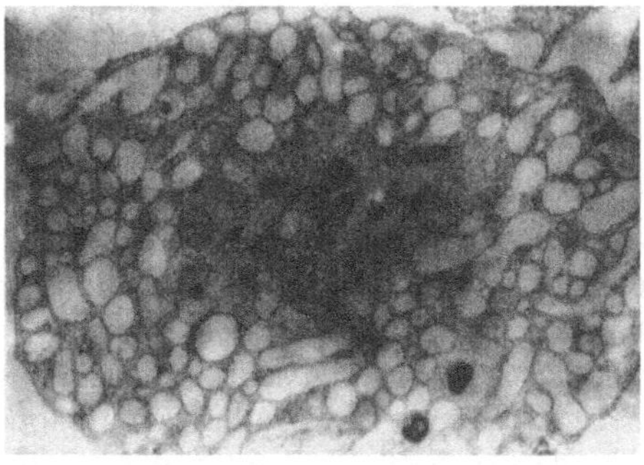

Figure 87. Porous structure of cytoplasm resulting from widening cavities of smooth endoplasmatic reticulum 24 h after PRP-1 administration (5 x 10^{-3} M) (electron microscopy studies).

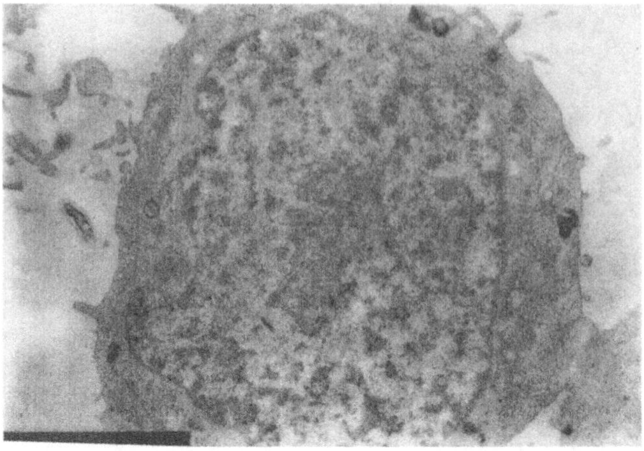

Figure 88. Electron-microscopic pattern of a NRGG-1 cell without of PRP-1.

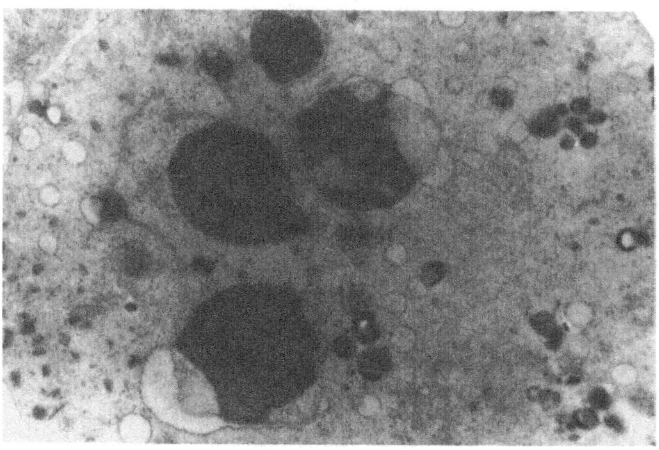

Figure 89. Apoptotic changes in NRGG-1 cells under PRP-1 action (5 x 10^{-5} M).

The electron microscopic data show the destruction of apoptosis type both in nervous tumor and L_{929}[141]. Two questions are raised: what are the differences between biochemically and genetically differentiated and non-differentiated neuroblastomas and why does PRP-1 produce the neuroprotective effect on the SC cells, but destruct tumor cells (*NRNG* and L_{929} fibroblasts)? Apparently, PRP-1 can affect the viability of different lymphocytes and other cells at different stages of differentiation and in various functional states. This is confirmed by data on PRP-1 influence on the activity of blood lymphocytes from different donors. PRP-1 also possesses properties of NGF, affecting both neurons and glia, including the stimulation of GFAP synthesis in oligodendrocytes and the formation of ILs in astrocytes. Transformed cell lines like the malignant cells isolated from tumors are more sensitive to pro-apoptotic stimuli than normal cells.

Conclusions. It was difficult to observe morphological changes in PRP-1-treated transformed cells by light microscopy. However, a combination of light and electron microscopy enabled us to find the following changes in cell proliferative activity and morphology:

a) PRP-1 exerts weak mitogenic effect on *NRNG* cells not provoking additional forms of pathological mitoses, which is characteristic of this tumor cell line.

b) PRP-1 induces a comparatively low increase in the number of gigantic cells in *NRNG* culture, which provides evidence of dysfunction of cytotomy.

c) PRP-1 at a concentration of 5×10^{-3} M exerts cytotoxic effect on certain tumor cells, which is manifested by cell destruction and apoptosis (electron microscopy studies).

7.2. CHANGES IN THE ULTRASTRUCTURE OF L_{929} TUMOR CELLS UNDER THE ACTION OF PRP-1 *IN VITRO*

The aim of the next part of our work was to study the effect of PRP-1 on the ultrastructure of cells of transformed mouse fibroblast line L_{929}. The experiments were carried out on 3-day cultures. The cultures were grown in RPMI-1640 medium with 10% FCS supplemented with 2 mM glutamine and 50 μg/ml of gentamycin, in a CO_2 incubator. The following variants were studied:

I - L_{929} cells cultured without PRP-1 (control); II - L_{929} cells cultured in the presence of PRP-1 at a concentration of 10^{-5} M; III – 10^{-6} M of PRP-1; IV - 10^{-6} M of PRP-1.

Cells were taken for the electron microscopy 24 hours after PRP-1 addition. After pelleting, the cells were fixed with 2.5% glutaraldehyde on the cacodylate buffer (pH 7.3) with post-fixation with 1% OsO_4 in the same buffer and embedded in eponym, araldite or westopal W. The experiments were repeated three times. Fixed ultrathin slices were studied by electron microscope (Galoyan et al., Medical Science of Armenia, 41:25-29). Besides, the mitotic activities were studied in control and experimental cultures. The mitotic activity was 80% in control; after the action of PRP-1 the total amount of mitoses in cultures II, III, and IV was 20.8%, of which K-mitoses amounted to 14%. Under the action of PRP-1 the ultrastructures of mature and immature L_{929} cells underwent changes (Fig. 92) in comparison to control (Figs. 90, 91). For mature cells these changes were the following: significant enlargement of non-granular cytoplasmic reticulum, fragmentation of nuclei and cytoplasm, destruction of the internal membranes of swollen mitochondria (Fig. 93). As for immature cells, they underwent other changes in the ultrastructure: pyknotic nuclei appeared (Fig. 94), as well as nuclei with karyolysis and karyorrhexis, and increase in microplasmatosis was observed. There were no changes in granular and non-granular cytoplasmic reticulum. It was of interest that large amount of mono- and polyribosomes was present in the cytoplasm of mature and immature L_{929} cells. Upon addition of PRP-1 to L_{929} culture (10^{-6} M and 10^{-7} M) a phenomenon of microplasmatosis took place in most cases. Numerous mono- and polysomes and single ruptured mitochondria were observed in the cytoplasm, but in general, the mitochondrial apparatus was maintained. The intergranular area remained enlarged. Vesicles of various sizes (from 50 to 200 nm and even more), fusing with each other to form larger vesicles and vacuoles, were observed in the cytoplasm on the periphery of the cells (Fig. 95).

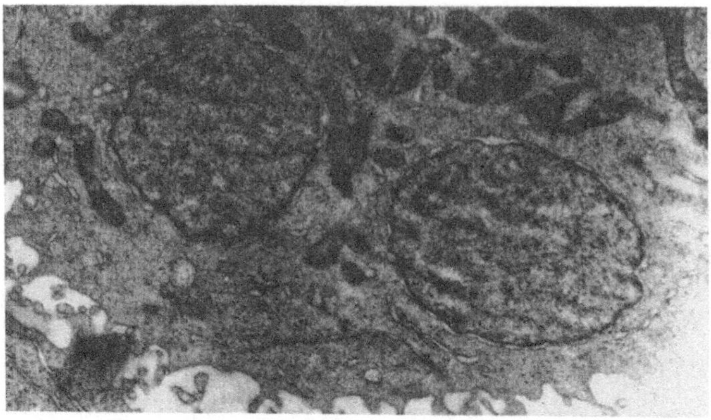

Figure 90. Three-day double-nuclei tumour cells in L_{929} monolayer culture. Control. Numerous mitochondria and cisterns of granular and non-granular reticulum are seen. Cytoplasm structure is poorly expressed. Mono- and polyribosomes are visible. Magnification x 6,000.

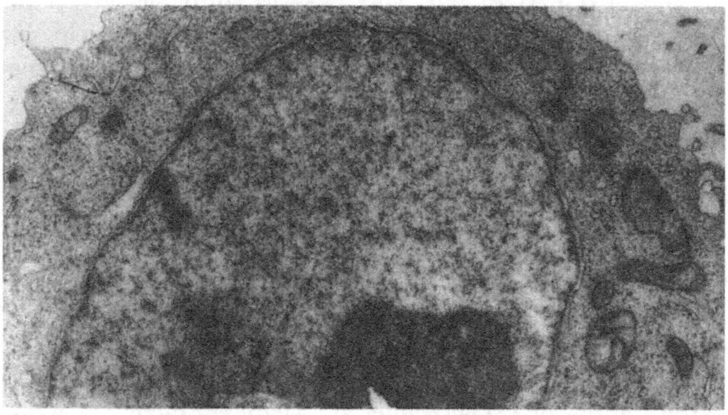

Figure 91. L_{929} tumor cell (fragment), control. Nucleus, nucleolus, cytoplasm, mature and immature mitochondria, mono- and polyribosomes are seen. Magnification x 6,000.

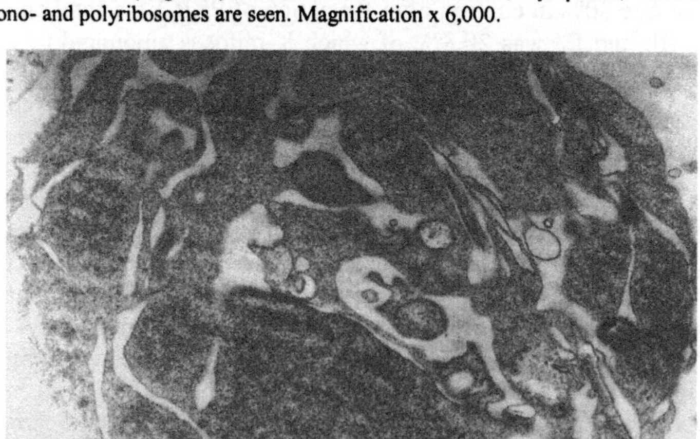

Figure 92. Mature L_{929} cell in monolayer culture after PRP-1 administration (10^{-5} M). Fragmentation of nucleus and cytoplasm is seen. Magnification x 10,000.

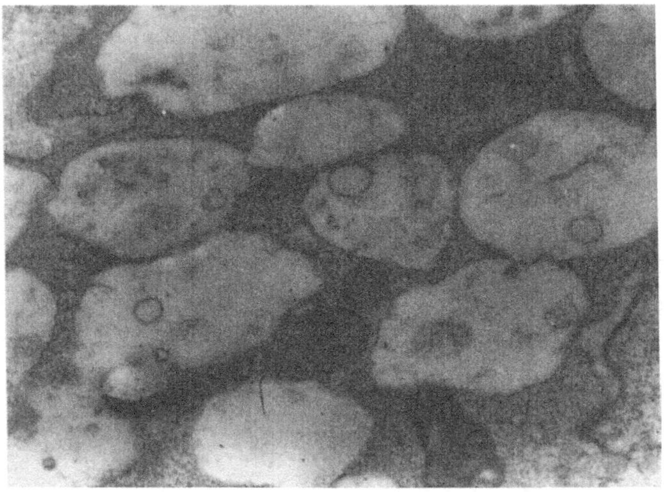

Figure 93. Effect of PRP-1 on 3-day monolayer culture of L_{929} cells (24 h after the peptide addition at a concentration of 10^{-5} M). Ruptured and swollen mitochondria in the cytoplasm are seen. Magnification x 20, 000.

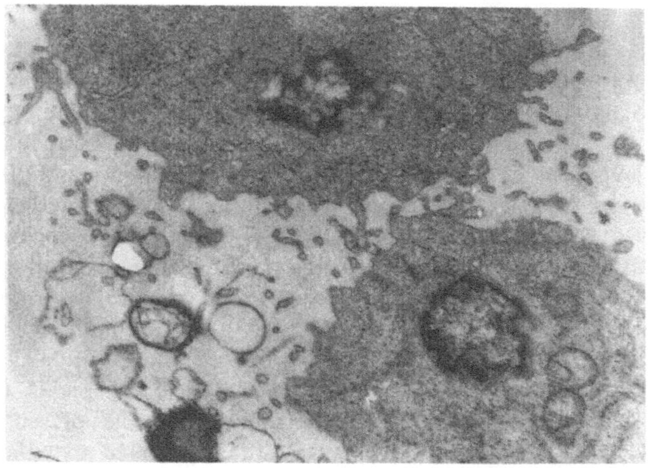

Figure 94. Immature 3-days L_{929} tumour cells after treatment with PRP-1 in concentration 10^{-5} M. Pyknotic changes of the nucleus. Increased microplasmatosis. Single mitochondria. Detritus of the decayed cells. Magnification x 6,000.

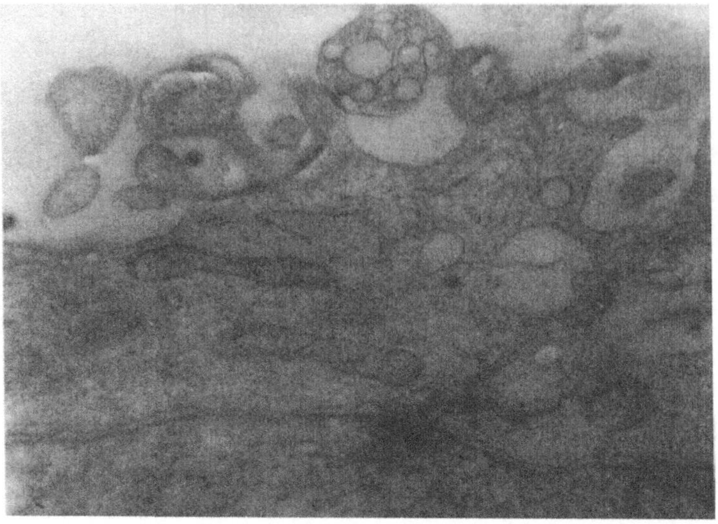

Figure 95. Effect of PRP-1 on L_{929} 3-day cell monolayer, 24 h after the peptide addition (10^{-6} M). Microplasmatosis, microvesicles of various diameters, and mitochondria are seen. Magnification x 20,000.

In the meantime, the mitotic activity of control and experimental cultures was calculated. In control it was 80%, while in the presence of PRP-1 in cultures II, III, and IV it was 20.8%. In culture II the inhibition of mitotic activity was observed. It was of interest that K-mitoses were observed in cultures II, III, and IV, which in total accounted for 14%. According to our data, the most efficient proliferation-inhibiting concentration of PRP-1 was 10^{-5} M. Data obtained from the analysis of ultrastructure of L_{929} cells grown in monolayer provide evidence that PRP-1 at a concentration of 10^{-5} M causes the decay of mature and immature tumor cells and significant changes in their ultrastructural morphology. The plasmalemma also undergoes destruction, a phenomenon called microplasmatosis. The synthesis of cytoplasmic proteins increases as judged from mono- and polysome content. Proliferation of the cells is accompanied with their decay.Thus, the decay of cells is clearly observed after 24-hour cultivation in the presence of PRP-1. The decay of mature cells is manifested by the fragmentation of nuclei and cytoplasm, while that of immature cells results in a sudden decay of nuclei (karyopyknosis, karyorrheksis, and karyolysis). In the cells cultured in the presence high PRP-1 concentration (10^{-5} M), ruptured membranes and fragmented cytoplasmic processes occur.

Thus, it was established in our experiments that PRP-1 at a concentration of 10^{-5} M induced the decay of tumor cells L_{929} and decrease in their mitotic activity and caused K-mitoses (pathologic mitotic fission). PRP-1 concentrations of 10^{-6} M and 10^{-7} M were less effective.

8

BASIC BIOCHEMICAL MECHANISMS OF PRP ACTION IN DIFFERENT PATHOLOGIES

8.1. EFFECT OF PRP-1 ON GFAP BIOSYNTHESIS IN ASTROCYTE CULTURE (IMMUNOCYTOCHEMICAL DATA)

Four structurally and functionally related proteins have been identified within the family of neurotrophins. These are NGF[280], brain derived neurotrophic factor (BDNF)[281,282], neurotrophin-3 (NT-3)[283,284], and neurotrophin 4/5 (NT-4/5)[285,286]. All neurotrophins bind to a common low-affinity neurotrophin receptor and have both distinct and overlapping effects on specific high-affinity receptors belonging to Trk family of tyrosine kinases: TrkA, TrkB, and TrkC. While TrkC can only bind NT-3 with high affinity, TrkB binds BDNF and NT-4/5 with high and NT-3 with lower affinity[287]. Biological effects of neurotrophins include stimulation of survival, growth, differentiation, and maintenance of neurons. Interest in neurotrophic factors grew in recently connection with their potential use in clinical practice for the correction of metabolic processes, as well as for specific protection of neurons in the therapy of neurodegenerative diseases, brain ischemia, and stroke[288-290].

In regard to possible neurotrophin effect on the cell systems of human embryonic neural tissue, there are only a few articles demonstrating the effects of recombinant human neurotrophic factors on free-floating roller tube cell cultures of human embryonic ventral mesencephalon. At the same time, BDNF effect on the cells of embryonic ventral mesencephalon was estimated, and the levels of dopamine and serotonin were measured as well as the activities of glutamate decarboxylase and tyrosine hydroxylase.

To our knowledge, there are no data in the literature on the effect of neurotrophic factors on metabolic processes in glial cells. A few reports about the promoting effect of neurotrophic factors on cell cultures of astrocytes, oligodendrocytes, microglia, and Schwann cell were published[287-296].

Recently we published a report about the promoting effect of BDNF, GDNF, and PRP-1 on GFAP biosynthesis in astrocytes of rat embryonic neural tissue[297].

Rat embryonic astrocytes were obtained from cerebellum of rat embryos on 13th day of gestation. Eight Wistar pregnant rats were decapitated under ether narcosis. Uteri with embryos were isolated under sterile conditions and placed into 70% ethanol for 10 min. After that, the embryos were removed from the uteri and transferred into the Hanks medium (Gibco) without antibiotics. Embryonic brain was isolated from the cranium and cleared of meninges and vessels. The cerebellum was separated from the brain, placed into Hanks medium (Gibco) without antibiotics and washed twice. The cerebellar tissue was cut into small pieces (5 - 10 mm^3), which were transferred into 0.25% trypsin solution (Sigma), volume ratio tissue/solution being 1:3. After 10-min incubation at 37°C the mixture was pipetted using 10-ml pipette for 2 min which was sufficient for obtaining a homogeneous suspension. The trypsinization was stopped by RPMI 1640 medium (Gibco) supplemented with 10% FCS. Test tube with cerebellar cell suspension was kept upright for 5 min. for cell sedimentation. The supernatant was removed, and the sediment was washed 3 times with Hanks medium (Gibco) without antibiotics. Then supernatant was centrifuged at 1000 rpm for 4 min. Then it was removed, and the sediment was washed with Hanks medium (3 times). After the third washing, the cell sediment was suspended in 1.0 ml of RPMI 1640 medium (Gibco) supplemented with 10% FCS. The viability of dissociated neurons was evaluated with trypan blue dye. The cells were resuspended in RPMI 1640 supplemented with 10% FCS and dispensed into the wells of three 24-well Costar plates precoated with poly-L-lysine (5 μg/ml).

For studying trophic factor effect the cultures with cell population density of 2 - 2.5 x 10^5 cells/cm^2 were used. The cells were cultured for 35 - 50 days. Astrocyte growth was detected after 12 days of culturing. The cultured of rat embryonic astrocytes were fixed in the mixture formalin : acetic acid : ethanol = 1 : 2 : 7 for 30 minutes. The cells were washed with 0.05 M phosphate buffered saline (PBS) three times. One ml of 5% BSA was added to the wells with rat embryonic astrocytes, and the plates were kept at 20°C for 30 minutes, then the cells were washed with PBS three times. FITC-conjugated rabbit antibodies to mouse immunoglobulins were used for visualization, and the incubation lasted two hours at 20°C. The microscopy was performed with "Leica" fluorescent microscope (magnification 1 × 100).

Immunoenzyme analysis of GFAP. GFAP produced by astrocytes was determined in culture supernatant by a sandwich immunoenzyme assay with monoclonal anti-GFAP antibodies. The procedure of GFAP immunoenzyme analysis was described elsewhere[297].

Description of the experiment. The supernatant (100 μl) from the wells containing 36-day old rat astrocyte culture was taken for quantitative GFAP analysis.

BDNF, GDNF, and PRP-1 in concentrations of 0, 10, 50 and 100 ng/ml each were then added to the cells, and the effect of neurotrophins on GFAP biosynthesis was analyzed 14 days after neurotrophine administration.

Results. The results of light and fluorescent microscopy of 36-day old rat embryonic astrocyte culture are shown in Fig. 96. As is clear from the figure, specific immunofluorescent staining with anti-GFAP antibodies makes it possible to identify the astrocytes producing this protein in culture.

Immunoenzyme test-system for GFAP with a sensitivity limit of 1 ng/ml (the calibration curve is shown in Fig. 97) was used to quantify the antigen in astrocytes culture medium. In this case the average GFAP concentration was determined as 28.6 ± 3.2 ng/ml.

The effect of neurotrophins on the biosynthesis of GFAP was studied in 14 days after BDNF, GDNF, or PRP-1 addition to the culture medium.The results are shown in Fig. 98. As is clear from the figure, all the neurotrophins under investigation were capable of

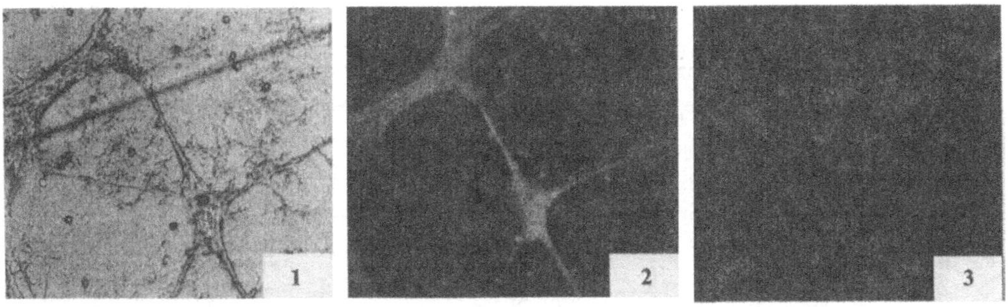

Figure 96. Immunofluorescent analysis of GFAP in rat embryonic astrocyte culture. 1 – Light microscopy. 2 – Same, immunofluorescent microscopy,treatment with mouse monoclonal antibody to GFAP and rabbit anti-mouse FITC-IgG. 3 – Control, mouse anti-NSE as the primary antibody. Magnification x 100.

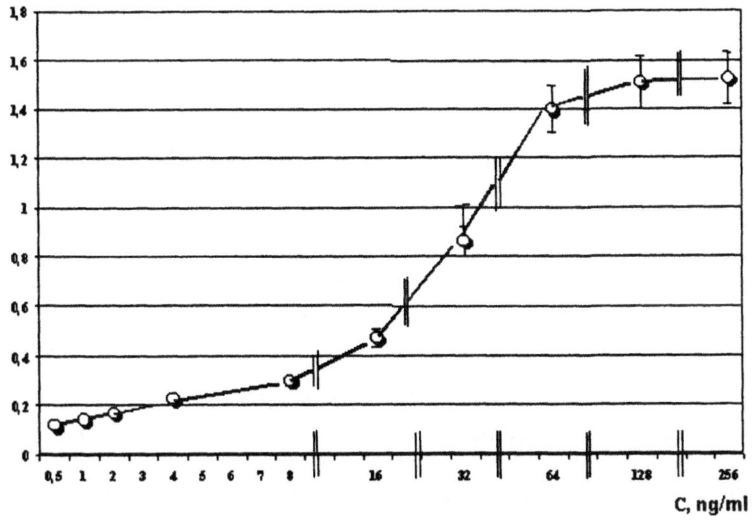

Figure 97. Calibration curve for GFAP determination by immunoenzyme analysis.

producing a stimulating effect on the synthesis of GFAP by embryonic astrocytes.

It should be pointed out that GDNF is the most active stimulant of GFAP biosynthesis, because 10 ng/ml of GDNF was enough to produce reliable increase in GFAP level in culture supernatant. However, similar concentrations of BDNF and PRP-1 did not induce reliable activation of GFAP synthesis. At the same time, concentrations of 50 and 100 ng/ml of all the neurotrophins under investigation were able to cause a valid stimulation of GFAP synthesis in rat embryonic astrocytes.

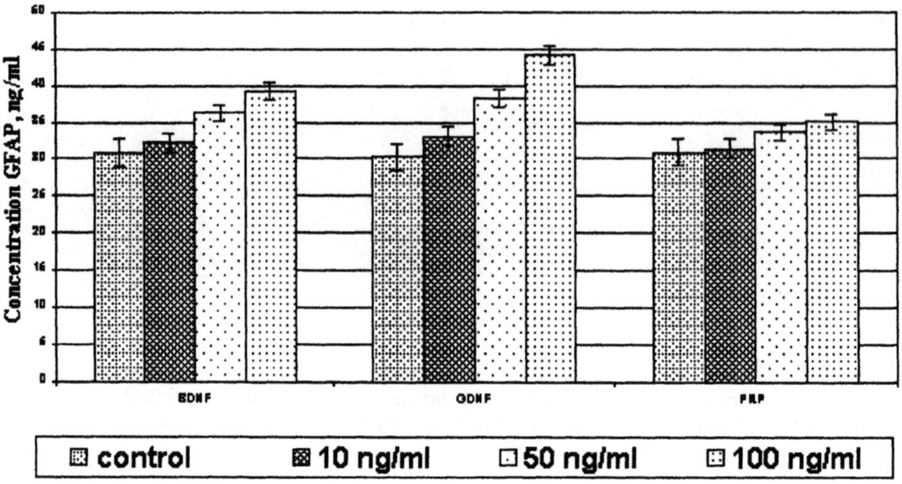

Figure 98. Effect of neurotrophin on the synthesis of GFAP in rat embryonic astrocyte culture.

It makes sense to introduce a special coeficient to compare the stimulating effect of neurotrophins on the biosynthesis of specific cellular proteins. CNS^{100} is a ratio of a concentration of a specific protein present in the culture fluid under the action of 100 ng/ml of a neurotrophin, to a concentration of this protein in the absence of the neurotrophin.

This coefficient makes it possible to compare stimulating effects of various neurotrophins. Thus, CNS^{100}_{BDNF} is 1.27 in our case, CNS^{100}_{GDNF} is 1.46, and CNS^{100}_{PRP} is 1.13. It is quite obvious that the promoting effect of GDNF on the synthesis of GFAP in rat embryonic astrocyte culture is significantly stronger than the effect of BDNF and PRP-1.

Summarizing, we conclude that BDNF, GDNF, and PRP-1 neurotrophic factors are capable of producing a stimulating effect on the biosynthesis of GFAP by astrocytes. Considering the effect of ultra-low concentrations of the neurotrophins, we may propose receptor-bound mechanisms of their action, which is in accordance with the data on the presence of TrkB-receptors on astrocytes.

In our view, the evaluation of neurotrophin effect on glial cells by quantitative determination of GFAP, an astrocytes-specific protein, may be recommended as a method to study the effect of any biologically active compound on glial cells. Further research into mechanisms of neurotrophin action on metabolic processes in embryonic astocytes may not only improve the therapeutic effect of glial cell transplantation, but also open new prospects in controlling glioses.

8.2. EFFECT OF PRP-1 ON CASPASE ACTIVITY OF MURINE NEUROBLASTOMA N2A

For the understanding of biochemical mechanisms of PRP-1 neuroprotectory effect during neurodegenerative changes in the nervous system, particularly in the brain, it is very important to study the activity of effector caspases, Asp-enzymes connected with neuronal survival, and mainly caspases that take part in the formation of amyloid-like polypeptides.

It is known that processing of amyloid precursor protein (APP) leads to the formation $A\beta_{1-42}$ polypeptide fragment which is the major etiological factor to account for neuronal loss and dysfunction in AD. $A\beta_x$ deposits occur also in other neurodegenerative diseases. This fragment results from sequential cleavage of APP by secretases releasing secreted ectodomains (sAPP or NTF) and cell-associated C-terminal fragments (CTFs). One of these, a 99-residue polypeptide CTF (C-99), contains $A\beta_x$ cleavage within its single hydrophobic transmembrane domain. This cleavage is made by uncharacterized endopeptidase (γ-secretase), and this results in the formation of Aβ fragments with variable C-termini. A family of novel Asp proteases (β-amyloid converting enzyme, β-secretase of BACE-1/2, memapsins) mediate the formation of C-99 from APP. Another secretase, α-, destroys the amyloidogenic domain of APP to release peptide C-83. While a specific α-secretase is not available, this may be done by a number of metalloendopeptidases including TNF-converting enzyme, or TACE which has putative properties of α-secretase[298].

The levels in cerebrospinal fluid (CSF) of β_{1-42} peptide and tau protein, the major constituents of amyloid plaques and neurofibrillary tangles, respectively, were proposed as biochemical markers of AD. Also, the amyloid plaques contain glycosamineglycans (GAGs), originating from chondroitin sulfate (ChS) and heparin sulfate (HS) proteoglycans, which play a pivotal role in amyloidogenesis[299]. The caspases, including caspase-3 and caspase-9, belong to cell-death effector proteases.

To examine the effect of PRP-1 further, we treated murine neuroblastoma N2A cells with the peptide in presence or absence of butyrate (Fig. 99), an agent known to induce both differentiation and apoptosis along with caspase activation in a variety of cell lines[300-304]. N2A provide a convenient model of a neural-like cell line where changes in death effector proteases (caspases) reflect cell viability[305]. The caspases act as an enzyme cascade causing the destruction of cytoskeletal and regulatory proteins[306].

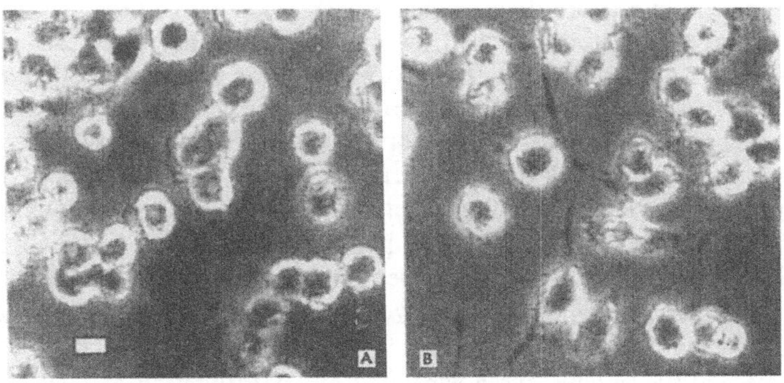

Figure 99. Phase contrast microscopy of murine N2A neuroblastoma cells grown in absence (A) or presence (B) of 10 mM butyrate. Note elongation and bipolar neurite extension in approximately 30% of cells in (B). Bar = 20μ.

In this study, we compared the effect of PRP-1 on putative proximal members of caspase family (caspases 2, 8, and 9) with changes in putative distal caspases (3 and 6), using preferred synthetic substrates, as described previously[307]. Data obtained show that PRP-1 affects cytosolic caspases in differentiated but not in non-differentiated N2A cells, providing another example of biological activity of this C-terminal proline rich NVAG fragment (Table 20).

Table 20. Effects of PRP-1 on caspase activity in murine N2A cells treated with butyrate

Caspase	24 hr		% change in presence of peptide at 24 hr
Proximal	-	+	
-2	22.82 ± 1.56	36.70 ± 2.42	61*
-8	25.19 ± 1.62	24.13 ± 1.22	-4
-9	12.80 ± 0.30	7.27 ± 0.81	-43*
Distal	-	+	
-3	107.76 ± 7.82	82.09 ± 0.08	-24*
-6	27.57 ± 4.99	55.43 ± 3.28	101*

Caspase members assayed in cytosolic extracts of N2A by hydrolysis of specific substrates as follows: Z-VDVAD-afc, Ac-IETD-amc, Z-LEND-afc, Ac-DEVD-amc, and Ac-VEID-amc for –2, -8, -9, -3, and –6 respectively as described in Methods. * - p 0.05, using 2-tailed Student's t-test; "-" - cells cultured in medium containing 10 mM butyrate; "+" - addition of 1 μM neurophysin-II fragment (PRP-1).

Cell Culture. Murine neuro2A (N2A) cells (a gift from S. Petanceska, Dementia Research Program, NKI) were grown in 1:1 DMEM/Opti-MEM medium (Gibco, Grand Island, NY, USA) containing 2% (V/V) FCS. Cells were passaged and grown to 70% confluence prior to all treatments. PRP-1 was dissolved in sterile DMSO at a concentration of 1 mM and added to cultures at a final concentration of 1 μM. Culture plates were placed in a humidified incubator with 5% CO_2. To induce differentiation, 10 mM sodium butyrate was added, and the cells were incubated for 24 - 72 h at 37°C. Culture medium was replaced every 24 h to remove non-adherent cells and to maintain pH and supply of nutrients depleted by rapid cellular metabolism and proliferation. Cells were examined by phase contrast microscopy as described previously[307-308].

Enzyme measurements. N2A cells were harvested, washed twice by ice-cold PBS, and homogenized in a Dounce homogenizer with the tight-fitting pestle with 75 strokes in 200 μl of 0.32 sucrose in 20 mM HEPES, pH 7.5, containing 30 mM KCl, 5 mM $MgCl_2$, 1 mM DTT, 1 mM 4-(2-aminoethylbenzenesulfonyl) fluoride (AEBSF) (Sigma), 10 μg/ml of aprotinin (Sigma), and 20 μg/ml of leupeptin and pepstatin (US Biochemicals, Cleveland, OH, USA).

Cytosols and membranes were separated by centrifugation at 16,000 g for 30 min and analyzed immediately or stored at –80°C. Aliquots of cytosols from 3 - 4 experiments were assayed for caspase activity as described previously with slight modifications. Proximal caspases 2, 8, and 9 were assayed with Z-VDVAD-amino-fluoro-coumarin (-afc), Ac-IETD-amino-methyl-coumarin (-amc), and Z-LEHD-afc, respectively, and distal caspasses 3 and 6, with Ac-DEVD- and Ac-VEID-amc (RBI, Natick, MA, USA, and

Calbiochem, La Jolla, CA, USA). Substrates were incubated at a final concentration of 50 μM in a volume of 50 μl for 75 min at 37°C, and reaction products were analyzed by RP-HPLC on a Vydac 218TP54 5 μC_{18} column. Free amc was eluted at a flow rate of 1 ml/min by 30% CH_3CN in 0.05% TFA, and elution of substrates containing afc was performed at 1 ml/min with a gradient of 0-60% CH_3CN in 0.05% TFA for 15 min to separate fluorophore from non-hydrolyzed substrate. Released fluorophore was detected at 365 nm excitation and 440 nm emission. Results were stated as ± SD, and significance was determined by 2-tailed Student's *t*-test. Protein concentrations were determined by microplate bicinchoninic acid (BCA) assay according to manufacturer's instruction.

Results and discussion.

N2A grown to 70% confluency appear as rounded cells devoid of neurites according to phase-contrast microscopy (Fig. 99A). There was no change in basal levels of cytosolic caspase-3 measured with Ac-DEVD-amc between 0 and 24 h, but a rapid 6-fold increase in the enzyme activity was observed within 72 h (Fig. 100) which was accompanied by an increase in number of non-adherent and non-viable cells. Administration of 1 μM synthetic PRP-1 to the culture medium prevented the increase in caspase-3 activity at the 24 h time point but did not change the overall 6-fold rise of cytosolic caspase-3 activity at 72 h. To examine the effect of PRP-1 following neuronal differentiation, N2A cells were treated with 10 mM butyrate for 24 h in the presence of the peptide. At this time point a marked lowering of cell growth rate in the presence of butyrate takes place, accompanied by 30% increase in the number of differentiated cells with distinct bipolar morphology (Fig. 99B), as compared to control (Fig. 99A).

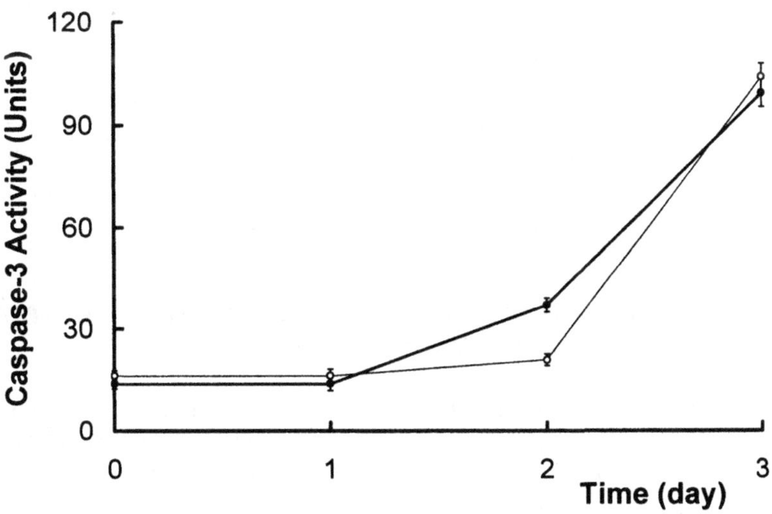

Figure 100. Caspase-3 activity measured in N2A cells grown up to 70% confluency for 3 d in the presence or absence of 1 μM of PRP-1.

In the framework of this model of cell differentiation, PRP-1 effect on caspases was different: there was a decrease in the activities of cytosolic caspases 3 and 9 (by 24 and 43%, respectively) which contrasted with an increase in caspase-2 and caspase-6 activities (by 61 and 101%); there was no significant effect on caspase-8 (Table 20). Our data demonstrated that PRP-1 significantly but in a different way affected several proximal and distal caspases in differentiated but not in undifferentiated N2A cells.

We have showed above that PRP-1 is a 15-mer peptide which amino acid sequence was identical to NVAG C-terminal fragment. This was established by sequencing of material purified from extracts of neurosecretory granules. The cells were able to differentiate by exposure to butyrate, and thus may be considered as an experimental model of neurons *in situ*.

Experiments with undifferentiated N2A cells showed a significant activation of caspase-3 as judged from the hydrolysis of the preferred substrate, Ac-DEVD-amc.

It was shown that distal caspases are upregulated in the cells of peripheral nervous system possessing THFα/Fas receptors, by conversion of proenzymes to form a cascade. Pathways mediating neuronal cell death, however, remain to be mapped[306]. The lack of THFα/Fas receptors on mature neurons suggests that alternative pathways account for the conversion of proximal caspases such as procaspase-9 by formation of a ternary complex consisting of mitochondrial cytochrome C and other intermediates[309]. Present data on N2A cells suggest that synthetic peptide PRP-1 does not target this intrinsic pathway since the activity of caspase-9 was reduced in its presence, which probably accounts also for decreased activity of caspase-3. Rather, the data suggest that PRP-1 acted on an alternative pathway for the conversion of procaspase-2, another proximal caspase, in a manner similar to the role this caspase plays in apoptosis of NGF-deprived PC-12 cells[310].

Activation of caspase-2 may account for the conversion of procaspase-6 and represents a pathway competing with Apaf-2 (cytochrome C)-mediated conversion of procaspase-9. A limitation of studies using synthetic substrates is a potential overlap of enzyme specificities, but it is impressive, nevertheless, that PRP-1 significantly increased the hydrolysis Z-VDVAD-afc rather than substrates for proximal caspases 8 and 9. Distal effector caspases such as caspase-3 and especially caspase-6, as shown here, probably dismantle cells by cleaving essential proteins such as laminin, Ca^{2+}-binding proteins, or ones linked to DNA integrity (PARP, ICAD), among others[306,311-313]. Our data on PRP-1, a peptide derived from C-terminal part of neurophysin-II precursor, indicate that its biological function is relevant to neuronal survival.

8.3. REGULATION OF MEMBRANE PHOSPHOLIPID METABOLISM BY PRP-1 IN CARDIOPULMONARY INSUFFICIENCY

Our experiments demonstrated that PRP-1 prevented complement-dependent hemolysis of sheep erythrocytes *in vitro*. These and other observations indicate that PRP-1 posseses a property to stabilize cell membranes, probably, by prevention of reactive radical formation, O_2^-, etc. The above mentioned data indicated that PRP-1 has a wide-range effect on the metabolism of membrane lipids, carbohydrates, and proteins. The obtained data suggest that PRP-1 has a wide metabolic effect in normal and pathological states. It was of interest to study enzymatic pathways of carbohydrate, lipid, and protein metabolism under PRP-1 action *in vivo* in normal condition and in experimental cardiopulmonary insufficiency[242-243].

Experiments were carried out on male rabbits weighing 2 - 3 kg. Cardiopulmonary insufficiency on the background of chronic inflammatory bronchopulmonary process was induced by violation of bronchial permeability by introduction of an alien body into the open space of trachea[243]. Six months after the operation synthetic hypothalamic PRP-1 preparation was injected into rabbits i/p at a dose of 25μg/kg for 3 days, then the rabbits were killed, and a morphological investigation was performed. Also we performed analyses of blood taken from 20 patients of the Department of Pulmonology of the Scientific Medical Center of the Municipal Emergency Hospital.

Fractionation of the individual phospholipids (PL) was performed by thin-layer chromatography on silicagel LC 5/40μ. After that the lipid phosphorus was determined[243]. Lipid peroxidation in tissues was studied by the reaction of malonic dialdehyde with tiobarbituric acid[314], and in the blood by a method described elsewhere[315]. Phospholipase A_2 activity was measured by spectrophotometric assay[316]. The same indices of blood erythrocytes and lymphocytes were counted. Determination of the listed biochemical indices was made at pre- and post-treatment periods. Determination of enzymes of PL biosynthesis such as glycerokinase (GK) and α-glycerophosphate dehydrogenase (α-GPD) was described earlier[317]. Dihydroxyacetone phosphate (DOAP) reduction was described elsewhere[318].

It was established that 6 months after inducing the chronic inflammatory bronchopulmonary process the cardiopulmonary insufficiency begins to develop.

As is characteristic of such type of disorder, essential disturbance of qualitative and quantitative content of almost all membrane PL, mainly phosphatides, was observed (Fig. 101). Under these conditions a pronounced increase in cytotoxic lysophosphatidylcholines (LPC) was observed along with pronounced decrease in phosphatidylcholines (PC), which testifies to phospholipase A_2 activation. Thus, cardiopulmanory insufficiency is characterized by noticeable increase in the content of sphingomyelins (SM) and phosphatidylinosites (PI), while cardiolipins (CL) content decreases (Fig. 101). Those changes appear on the background of some decrease ($p < 0.01$) in total PL content in the tissue of rabbit right ventricle.

Similar disturbance is observed (though less pronounced) in the case of erythrocyte membranes of experimental animals (Fig. 102). The increase in SPM content ($p < 0.001$) and decrease in that of PC, PE, and phosphatidylserine (PS) are more pronounced in comparison with normal state. Qualitative and quantitative content of PL from lung tissue of experimental animals is shown in Fig. 103.

These observations show significant changes in PL copmposition, which are manifested by sharp increase in LPC level with simultaneous (about two fold) decrease in PC, PE, and PG, which are more noticeable in the case PL of lung tissue. Changes in other PL fractions occur in different directions and are less expressed. An exception is SPM which the level is increased two-fold in pathologically changed lung tissue. The qualitative and quantitative changes in the content of PL in right heart ventricle, lungs, and erythrocytes under the action of PRP-1 are illustrated by Figs. 101-103. These results demonstrate strong effect of PRP-1 on PL metabolism.

In pathologically changed myocardium positive changes in the content of main classes of phosphoglycerides (PC and LPC) occur, and almost full normalization of PI, SPM, and CL level takes place. As known, the latter are localized predominantly in the mitochondria and are also main structural components of cardiomyocyte plasmatic membranes, which have important role in cardiac muscle activity.

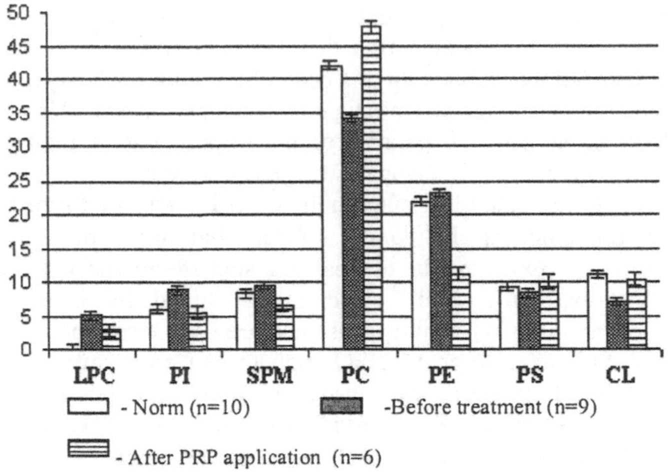

Figure 101. Content of different fractions of phospholipids (% of total) in tissues of the heart left ventricle in normal state, in cardio-pulmonary insufficiency, and after PRP-1 application.

PRP-1 regulates membrane phospholipides metabolism normalizing the functional activity of membranes. This provides evidence about certain normalization of LPC level (cytotoxic and membranolytic compounds). As to a strong change in PE content in diseased tissue, we suppose that it results from its conversion to PC and PS, metabolically more active compounds. More profound analysis made for clarification of the specificity and the degree of changes in phosphatide composition allowed us to find a number of principally new and, in our opinion, interesting regularities indicating the direction and high efficiency of hypothalamic PRP-1 action. It is worth noting that after PRP-1 administration the correction of PL content in cardiomyocyte membranes is accompanied by simultaneous normalization of total PL level in the myocardium. Our data indicate certain normalization of qualitative and quantitative composition of PL in erythrocyte membranes (Fig. 102) and in the cells of lung tissue (Fig. 103) of experimental animals after administration of PRP-1. Full correction of the level of cytotoxic LPC and PE, both in erythrocytes and in lung tissue, is of our special interest. A clear tendency to normalization of the content of other PL fractions, particularly SPM, PC, and CL, is observed under these conditions.

The increase in PI level (namely, monophosphoinositides (MPI)) under conditions of immune reaction changes in the organism, depends, in our opinion, on stimulation by these substances of biosynthic pathways compensating a decrease in di- and triphosphoinositides, which are important components of phosphoinositide signal system, transductors of exterior signal inside the cell.

It was of great interest to characterize changes in components of PI signal system in our further experiments. It is obvious from data shown in Figs. 104 and 105 that the development of cardiopulmonary insufficiency is characterized by statistically significant

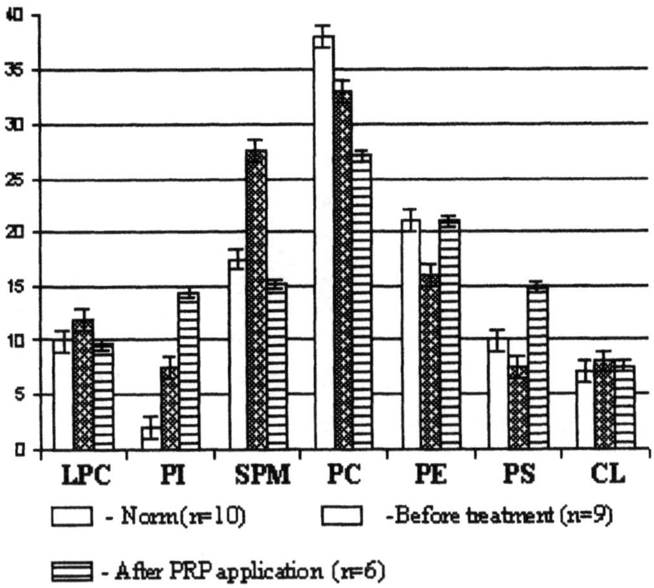

Figure 102. Content of different fractions of phospholipids (% of total) in rabbit erythrocyte membranes in normal state, in cardio-pulmonary insufficiency, and after PRP-1 application.

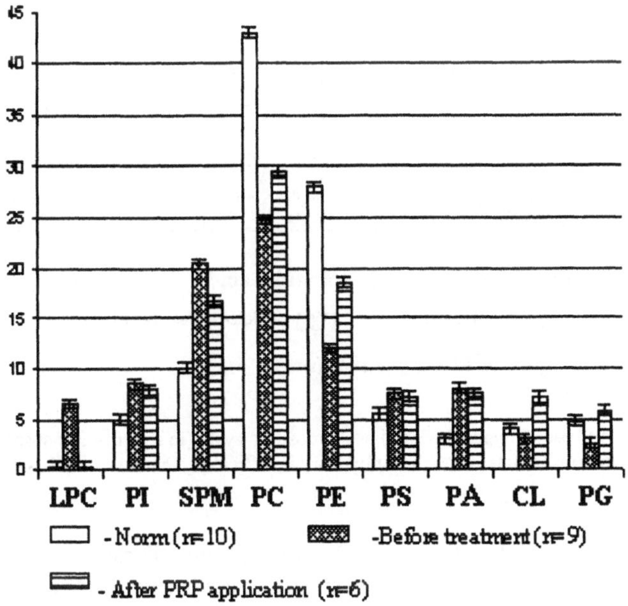

Figure 103. Content of different fractions of phospholipids (% of total) in rabbits lung tissue in norm, in cardio-pulmonary insufficiency, and after PRP-1 application.

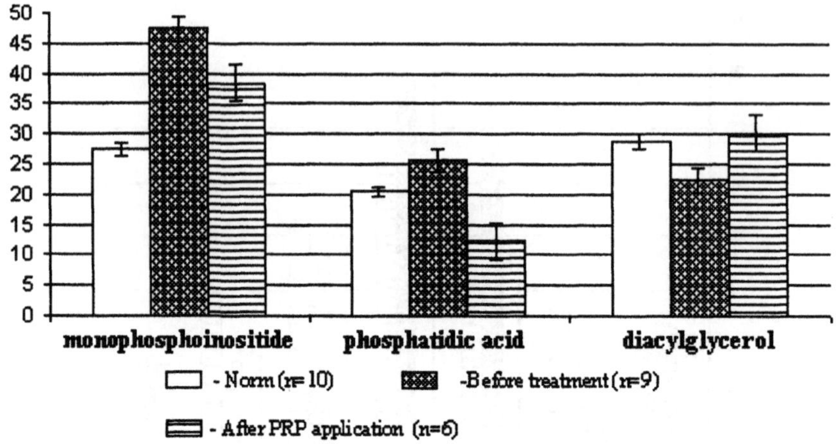

Figure 104. Level of different components of phosphoinositide cycle (in μg of lipid phosphorus per 1 g of fresh tissue) in rabbit heart right ventricle in normal state, in cardio-pulmonary insufficiency, and after PRP-1 application.

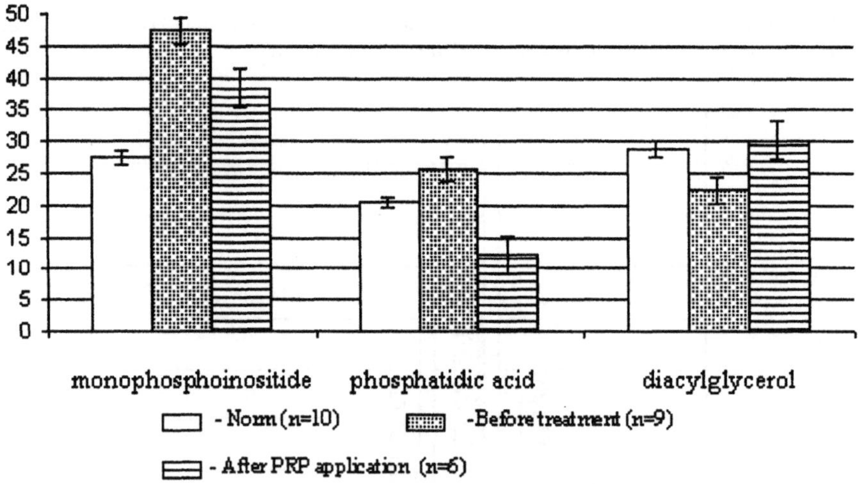

Figure 105. Level of different components of phosphoinositide cycle (in μg of lipid phosphorus per 1 g of fresh tissue) in rabbit lung in normal state, in cardio-pulmonary insufficiency, and after PRP-1 application

($p < 0.01$) changes in the level of PI cycle components. These changes are particularly well expressed in the case of MPI and phosphatidic acids (increase) and CL (decrease), in heart right ventricle and in lungs of experimental rabbits. Administration of hypothalamic PRP-1 results in the normalization of MPI as well as of phosphatidic acids and CL, which are secondary messengers of phosphoinositide signal system. PRP-1-induced changes are more pronounced in rabbit lung under pathological conditions (Fig. 105). These data enable to better understand the biochemical mechanisms of PRP-1 action on the immune system activity.

It seems to us that one of the essential pathways for formation of immunological reactions of the organism to be considered is the effect of hypothalamic PRP-1 on the functional state of certain receptors, the activity of phospholipases, and perhaps on the activity of protein kinases and regulatory systems of calcium ions transport via cell plasmatic membranes[319].

Our further studies aimed to investigate the molecular mechanisms of qualitative and quantitative changes of membrane PL in cardiopulmonary insufficiency and the effect of hypothalamic PRP-1 on these changes. The results obtained (Fig. 106) demonstrated profound disturbances in the activity of enzymatic systems at initial stages of PL biosynthesis, namely in GK and GPD (glycerophosphate dehydrogenase). Development of cardiopulmonary insufficiency induces some inhibition of GK activity both in myocardium (32%) and in lungs (94%), which indicates a decrease in the rate of formation (via GK pathway) of phosphoglycerides in these organs of the experimental animals.

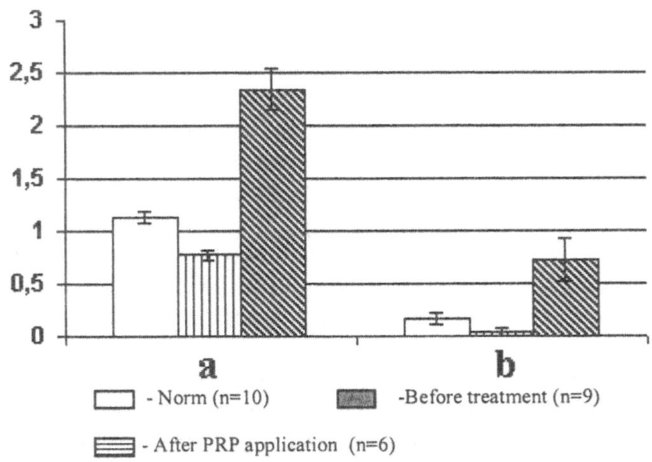

Figure 106. Activity of glycerokinase (GK) in tissues of rabbit heart right ventricle (a) and lung (b) in normal state, in cardio-pulmonary insufficiency, and after PRP-1 application.

Our investigations also demonstrate significant inhibition of the rate of DOAP reduction to form GP, i.e. the intensity of glycolytic pathways of PL biosynthesis in pathologically changed tissues, in heart right ventricle (55%) and lungs (49%). Inhibition of GK and GPD is accompanied by simultaneous decrease in GP quantity, which is a key metabolite for glycerolipid biosynthesis in myocardium (30%) and lung tissue (93%). The administration of hypothalamic PRP-1 leads to some correction of disturbed phosphatidogenesis which is characterized by a sharp increase in GK and GPD in the tissues of the heart right ventricle and lungs. According to our data the intensification of GK pathway is the most essential for GP formation in the studied tissues. Under these conditions a strong stimulation of GPD activity takes place in the organism. After administration of PRP-1 an obvious normalization of GP level in the studied tissues occurs (Fig. 107), the strongest positive effect being observed in heart right ventricle, which is well reflected by changes in glycerophospholipids. Comparison of activities of GK and GPD and cellular level of GP let us assume that the normalization of phosphatide-glyceride content in studied tissues depends on the increase in the rate of glycolytic and GK pathways of GP formation with further involvement of the latter in the reaction of phosphatidogenesis. We cannot exclude the possibility of inhibition of the processes in the case of degradation of PL membranes under the action of PRP-1, which needs detailed clarification. In our opinion, studying phospholipase activities is of certain interest for the understanding of their physiological role and significance for the patholody. Studying of phospholipase A_2 activity changes as well as comparative evaluation of changes in LPC composition in heart ventricle, lung, and erythrocyte membranes allowed us to predict a possibility of correction of membrane phospholipid composition in cardiopulmonary insufficiency. Data presented in Figs. 108 and 109 show the increase in phospholipase A_2 activity in the studied tissues, which induces inhibition of respiratory enzymatic system of cardiomyocites. It is known that one of results of phospholipase activity stimulation is the damage of biomembranes by hypoxic factors. That is why the efforts of scientists are directed to the search of powerful inhibitors of phospholipase A_2[322]. The effect of proteolytic enzymes on cellular membranes plays an important role in this process. In the presence of phospholipase A_2 and proteolytic enzymes complete lipolysis of PC and PS accompanied by further inhibition of respiratory enzymatic system occurs in various pathologies[320], including cardiopulmonary insufficiency[321]. Our data show that the use of the hypothalamic PRP-1 is accompanied by apparent inhibition of phospholipase A_2 activity in tissues of both heart right ventricle and lungs and in erythrocyte membranes (by 32.0, 53.0, and 14.7 %, respectively), as well as in lymphocytes of the experimental animals (Figs. 108 and 109), and of patients with cardiopulmonary insufficiency (Fig. 110). This indicates the inhibition of deacylation reactions, i.e. processes of membrane PL cleavage, in cardiopulmonary insufficiency.

Comparative evaluation of quantitative changes in the composition of PC, LPC, and other classes of phosphoglycerides of biomembranes, phospholipase A_2 activity, and enzymatic systems of initial stages of phosphatidogenesis, as well as the components of PI signal systems, in tissues and blood allowed let us to make a conclusion that one of possible pathways for regulatory and membrane-stabilizing effects of hypothalamic PRP-1 in cardiopulmonary insufficiency is the regulation of membrane lipids metabolism by changes of functional steroid receptors, mobilization of intracellular regulatory systems of Ca^{2+} release from subcellular structures under the action of phospholipases, increase in PL biosynthesis, and inhibition of lipolysis.

We obtained data on positive effect of PRP-1 on the level of lipid peroxidation (LPO) in heart right ventricle tissues, erythrocyte membranes, and lymphocytes, in cardiopulmonary insufficiency. It should be noted that a proline-rich cluster is present in the primary structure of phospholipase inhibitor Python (PIP) within 85-109 sequence (QCEDPLPGLPLSLQNGLYCPGAFG), which is probably an epitope for phospholipase A_2 inhibition. Interpreting the data obtained, it is important to take into consideration the fact that under experimental oxidative stress PRP-1 protects membranes by the increase in catalase activity. Thus, PRP-1 possesses antioxidant properties, which was

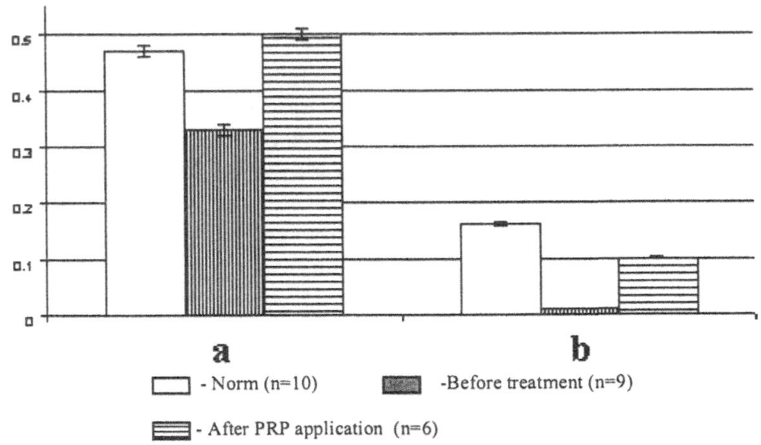

Figure 107. Level of glycerophosphate dehydrogenase (in μM of NADH/g of tissue) in heart right ventricle (a) and lungs (b) in normal state, in cardio-pulmonary insufficiency, and after PRP-1 application

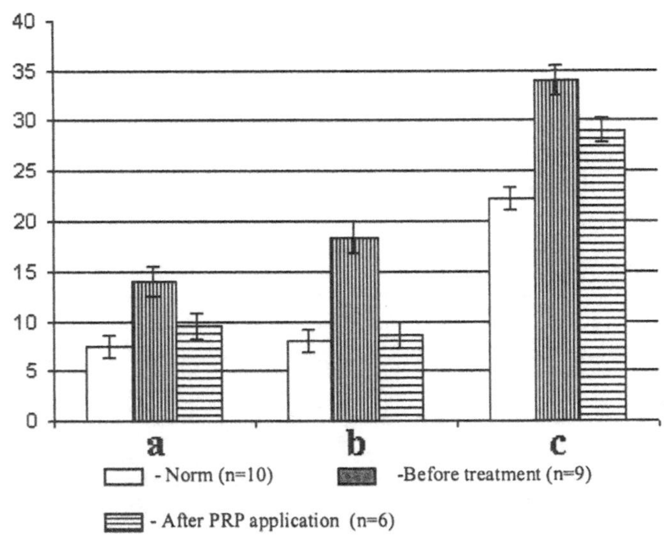

Figure 108. Activity of phospholipase A_2 in tissues of rabbit heart right ventricle (a), lungs (b), and erythrocyte membranes (c) in normal state, in cardio-pulmonary insufficiency, and after PRP-1 application.

demonstrated in our experiments: PRP-1 inhibited phospholipase A_2 activity and lowered the level of LPO products in cardiopulmonary disease. It is interesting to note that similar effect was observed in the case of erythrocyte membranes of blood cells, as well as tissues of heart and lungs, i.e. decrease in lysophosphatidylcholine formation. Prevention of H_2O_2 accumulation itself prevents the formation of oxygen active radicals (O_2^- , $HO^\cdot$, etc.) and preserves the membrane integrity. It is promoted also by GP level maintenance which is a key metabolite for glycerolipid biosynthesis and by increase in GK and GPD activities. In other words, PRP-1 takes an active part in phosphatidogenesis in different organs. This induces normalization of the content of monophosphoinositides, phosphatidic acids, and diacylglycerols (secondary messengers of cell signaling system).

Thus, our data show that the biological effect of PRP-1 is realized through the regulation of phosphatidogenesis and also through the modulation of myocardial cell activity by phospholipase regulation.

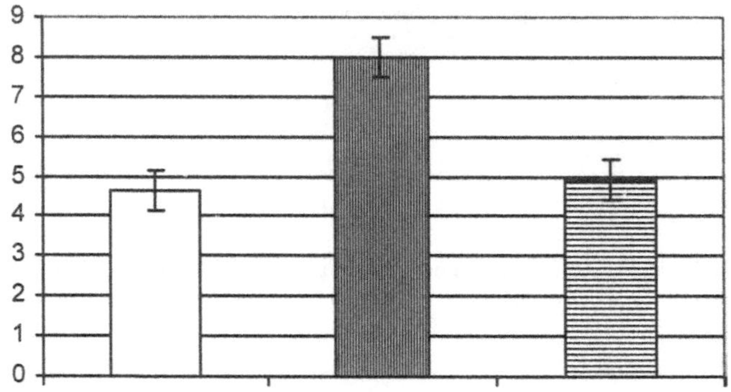

Figure 109. Activity of phospholipase A_2 in rabbit blood lymphocytes in normal state, in cardio-pulmonary insufficiency, and after PRP-1 application.

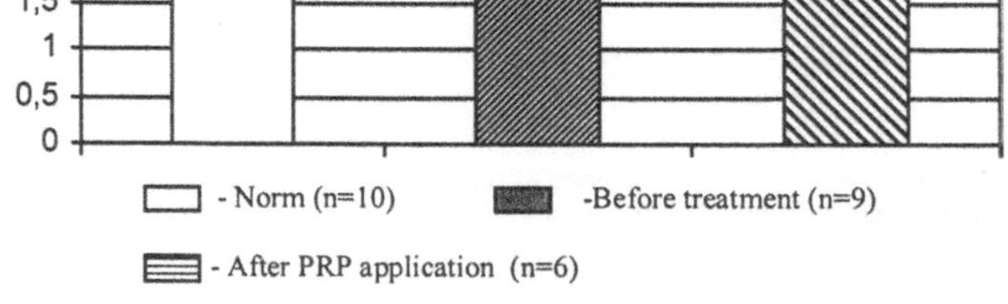

Figure 110. Activity of phospholipase A_2 in patients' blood lymphocytes in normal state, in cardio-pulmonary insufficiency, and after PRP-1 application (*in vitro*).

8.4. INFLUENCE OF PRP-1 ON CONSEQUENCES OF IONIZING IRRADIATION OF RATS

The aim of this part of our work was to study possible protective effect of PRP-1 against ionizing irradiation. We studied qualitative and quantitative changes in individual PL, changes in lipid peroxidation (LPO), and in activities of enzymatic systems of phosphatidogenesis, and also the degradation of PL of rat hepatocyte microsomal membranes.

Materials and Methods. Thirty two Wistar male rats (180 - 200 g) were used in the experiments.

The rats were treated with a dose of 3 Gy, radiation intensity was 0.26 Gy/min, the skin-focal distance was 60 cm. The animals were decapitated 10 days after the irradiation. Microsomal fractions of hepatocytes were received by the method of differential centrifugation at 17,000 g[318]. Fractionation of the individual PL was carried out by thin-layer chromatography on LC adsorbent. The activity of PL metabolizing enzymes GK and α-GPD was studied in the reaction of α-GP oxidation according to the method described elsewhere[317]; reaction of DOAP reduction was performed as described before[318]. Determination of LPO was published erlier[316], and the activity of phospholipase A_2 was measured by spectrophotometric method with PL as a substrate. Hypothalamic PRP-1 was injected once a day i/p, during 3 days, by 10 μg/kg of body weight. The significance of the data was assessed by Student's t-test.

Results and Discussion. Significant changes in the quantity of membrane hepatocyte microsomal PL occurred in 3 days after irradiation (Fig. 111). Under these conditions the concentration of LPC, which is a cytotoxic product of membrane PL degradation, increased almost 3 times, and PC concentration underwent a decrease greater than 2-fold. Statistically significant decrease ($p < 0.01$) in the concentrations of DPG, SM, PI, and MPI was observed. Simultaneously, a decrease in PE and accumulation of PA, which are intermediate products of phosphoglyceride biosynthesis, were observed. Administration of PRP-1 induced certain normalization of the composition of hepatocyte microsomal membrane PL: the content of LPC, PI, and PA was decreased and simultaneously PC level was significantly increased ($p < 0.01$, see Fig. 111). To find the concrete point of phosphatidogenesis disturbance, we focused on GC-catalyzed transformation of glycerol into glycerophosphate. Taking into account that dihydroxyacetone phosphate (DOAP), a product of glucose decomposition, participates in phospholipid biosynthesis, we studied also the activity of glycerophosphate dehydrogenase (GPD) in hepatocytes (Ghazaryan, Galoyan, Neurokhimia, 2003, in press). Substantial difference was found for reaction rates of DOAP reduction into α-GP and activation of free glycerol in the liver of control animals. The activity of GPD in the reaction of DOAP reduction into α-GP was approximately 10 times greater than the that of GK.

On the 10th day after irradiation the activity of α-GPD (which substrate is DOAP) significantly decreased ($p < 0.01$) on the background approximately complete loss of GK activity in hepatocytes (see Fig. 112). The inhibition of GK and glycolitic pathways of GP formation, which is a central metabolite of glycerolipid synthesis, are accompanied by significant decrease in GP content in the investigated tissues, and by simultaneous significant inhibition of mitochondrial GPD activity. The results obtained provide important evidence of decrease in of GK activity and inhibition of glycolytic pathways of biosynthesis of glycerolipides in the hepatocytes caused by ionizing irradiation. After PRP-1 administration almost complete normalization of GK activity was observed as well as obvious tendency towards normalization of both mitochondrial and cytoplasmatic α-GDP.

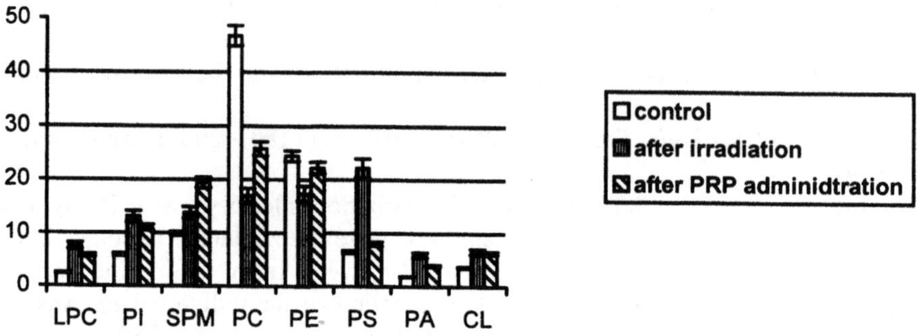

Figure 111. Changes in PL of hepatocyte microsomal membranes after irradiation and PRP application in comparison with normal state (%).

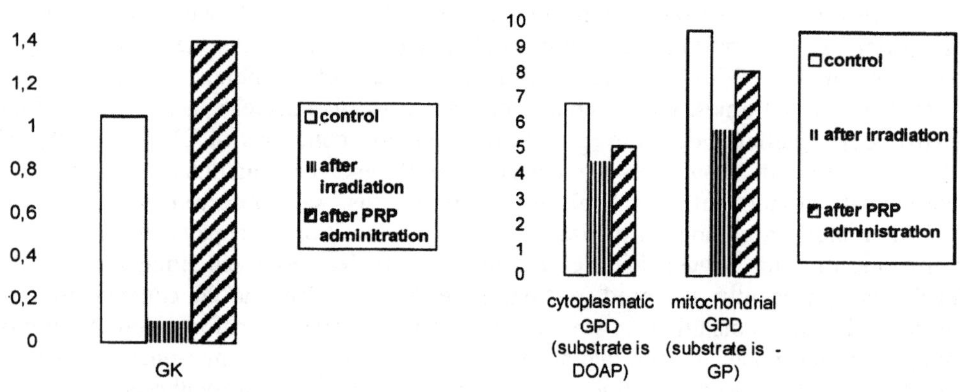

Figure 112. Activity of enzymes, which participate in the exchange of L- α-GP in hepatocytes in normal state and after PRP administration.

The activity of phospholipase A_2 and intensity of LPO were measured in hepatocytes of control irradiated rats and of those after PRP-1 administration (Fig. 113). It was established that the effect of ionizing irradiation is characterized by significant increase in the activities of both phospholipase A_2 and LPO. After administration of PRP-1 statistically significant ($p < 0.01$) decrease in phospholipase A_2 activity and in the rate of LPO was observed along with normalization of hepatocyte membrane function.

Thus, the injection of hypothalamic PRP-1 into rats after radiation exposure leads to a noticeable normalization of PL composition in microsomal membranes. PRP-1 exerts its regulatory effect on glycolytic and GK pathways of glycerophospholipid biosynthesis, on the one hand, and on the rates of primary stages of PL degradation and LPO, on the other.

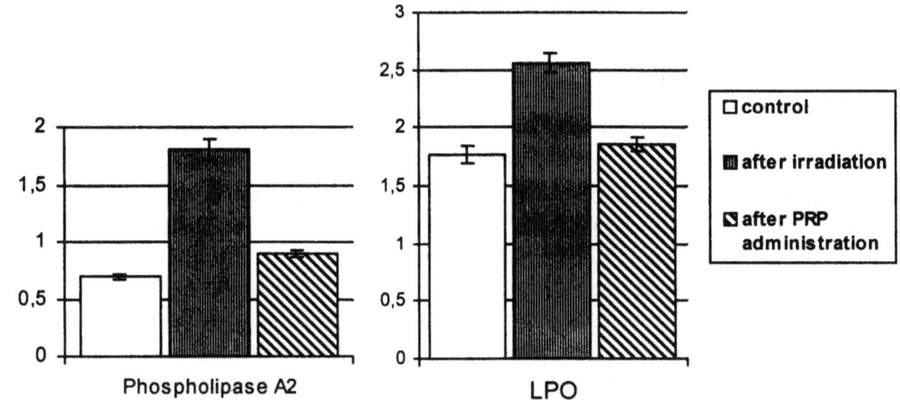

Figure 113. Changes of phospholipase A_2 activity (in standard units) and rates of LPO (μM of MDA/g) in hepatocytes in normal state, after ionizing irradiation, and after PRP administration.

DISCUSSION

The second part on the 20th century was marked by intensive development of neurobiological sciences, and especially of brain biochemistry. The neurochemical multistep regulatory mechanisms of endocrine gland function by hypothalamus were discovered; the mechanisms of synthesis and chemical structure of liberins, adenohypophysial hormones, were elucidated, as well as the molecular-genetic mechanisms of different neuroendocrine diseases. Nevertheless, one of the most striking mysteries of the brain and heart remained undiscovered: the existence of a large series of hypothalamus and heart neurosecretory neurohormones, multiple mechanisms of neurohumoral integration of neurosecretory hypothalamus with neurosecretory ganglionic cells of atria, also the role of neurosecretory cytokines and other signal molecules of the brain neuroendocrine immune system.

The biosynthesis of immune system signal molecules in the neurosecretory nuclei of hypothalamus (NPV, NSO) indicate that parvo- and magnocellular cells of these nuclei are immunocompetent cells. All experimental material described in the book confirms to a certain extent the existence of a neuroendocrine immune system of the brain, signal molecules of which participate in molecular mechanisms of immune response and neuronal survival.

The effect of recently discovered proline rich peptide PRP-1 is much broader - this cytokine is a unique mediator of the hypothalamus - neurohypophysis - bone marrow - thymus axis. PRP-1 may participates in hematopoietic stem cell differentiation. Our data demonstrated that PRP-1 is a stimulator of myeloid as well as lymphoid precursors. It regulates myeloid and lymphoid lineages differentiation.

PRP-1 is one of the neurotrophic brain factors. Four structurally and functionally related proteins have been identified within the family of neurotrophins: NGF[280], BDNF[281,282], NT-3[283,284], and NT-4/5[285,286]. Comparative study of PRP-1, BDNF, and GDNF influence on embryonic astrocytes from 13-day gestation embryos proved that there are grounds for making a conclusion that neurotrophic factors BDNF, GDNF, and PRP-1 are capable of producing a stimulating effect on the biosynthesis of GFAP by astrocytes. PRP-1 causes definite inhibition of caspases 3 and 9 and activates of caspases 2 and 6 in neuroblastoma N2A cells. As is known, these caspases cause degradation of some cellular proteins including precursors of amyloid peptides. There are literature data on positive role of caspase-2S in amyloid peptide formation. *In vitro* effect of PRP-1 on the activity of the abovementioned 4 caspases of neuroblastoma N2A cells indicates that *in vivo* PRP-1 probably prevents the development of neurodegenerative processes. Our data show that PRP-1 alters cytosolic caspases of differentiated but not of non-differentiated N2A cells providing another example of biological activity of PRP-1.

PRP-1 can be considered a stimulator of immune cells, myelopoiesis, T-cell differentiation, and hematopoiesis. Also, PRP-1 is a unique regulator of protein, carbohydrate, and lipid metabolism, with antioxidant action in different pathological conditions. PRP-1 can protect erythrocyte membrane during hemolysis preventing release of troponin C from cardiomyocytes into blood during experimental infarction (unpublished data). PRP-1 also inhibits membrane phospholipase A2 activity in rabbit heart tissue, erythrocyte membrane, blood lymphocytes, and cardiomyocytes during cardiopulmonary insufficiencies. It prevents of H_2O_2 accumulation and inhibits the formation of oxygen active radicals (O_2^-,$HO^{\cdot}$, etc.) in various toxicoses, thus maintaining cell membrane integrity. PRP-1 also maintains the level of glycerophosphate dehydrogenase and glycerol kinase activities, which are key metabolites in glycolipid biosynthesis. In other worlds, PRP-1 takes an active part in phosphatidogenesis, utilization of glucose, glycolytic pathways, etc. In Table 21 biological activities of PRP-1 are summarized.

Table 21. Biological properties of PRP

METABOLIC ACTIVITY IN PATHOLOGICAL CONDITIONS

• ⇓ phospholipase A_2; • normalizes PL metabolism;
• ⇓ lipid peroxydation; • ⇓ activity of caspase-3 and caspase-9; • ⇑ utilization of glucose;
• ⇑ protein biosynthesis; • ⇑ creatine kinase activity; • ⇑ activity caspase-2 and caspase–6; • ⇑ glycerol-3-phosphate kinase

ANTIBACTERIAL AND ANTIVIRAL EFFECTS

• ⇑ bactericidal activity of macrophages (phagocytosis, intracellular killing of bacteria, IL-1 synthesis, antigen-presenting function, accumulation and viability);
• ⇑ antibody production; • ⇑ survival of mice in several infections; • ⇓ growth of bacteria in internal organs and enhancement of their elimination from the body; • ⇓ *in vitro* replication of EMV in Hep-2 cells; • ⇑ interferon-γ level in mitogen-stimulated blood mononuclears.

MYELOGENESIS

• recovery of myelopoiesis in profound leukopenia induced by CPA in mice infected with *P.aeruginosa*

LYMPHOPOESIS
(T-cell differentiation)

• ⇑ growth and differentiation of mice fetal and neonatal thymocytes *in vivo* and *in vitro*

IMMUNOMODULATORY ACTION

• ⇑ immunocyte activation (T-cell proliferation; NK-cell, macrophage, monocyte, and neutrophil functions; astrocyte proliferation and expression of proinflammatory cytokines IL-1, IL-6); • ⇓ proliferation of Jurkat cells (human lymphoma cell culture) *in vitro*;
• ⇑ IL-2-dependent proliferation of T-lymphocytes, antibody-dependent T-cell and LAK-cell cytotoxicity; • ⇑ cytokine production (IL-1α, IL-1β, IL-2, IL-6, TNF-α)

NEUROPROTECTIVE PROPERTIES

• Restoration background and evoked neuronal activity of spinal cord inhibited by Vipera Raddei and Cobra venom intoxication;
• ⇑ regeneration after spinal cord hemisection; • neurotrophic effect

NEUROHORMONAL PROPERTIES

• ⇑ prolactin release into general circulation; • ⇑ expression of human GH in mice fibroblasts (clone BALB/c-GH).

A broad spectrum of PRP-1 biological effects on both immune and nervous system is accompanied by noticeable stimulation of carbohydrate, protein, and lipid metabolism. In the brain, both in normal state and in crush syndrome, a powerful PRP-1 stimulates glucose utilization. Thus, after 2-h decompression glucose utilization increases ~ 10 times under PRP-1 action. To a lesser extent the utilization of glucose is stimulated in other organs, such as heart, liver, and kidneys. High level of glucose utilization in the brain proves the PRP's role in carbohydrate metabolism, especially that of glucose. Simultaneous 3- to 4-fold increase in protein biosynthesis in subcellular compartments of the brain and noticeable regeneration of nervous tissue were shown.

PRP-1 is a potent antibacterial (*in vivo*) and antiviral (*in vitro*) agent. This property was realized through increase in macrophage and T–cell activity, stimulation of antigen presentation function of macrophages, antibody formation, etc.

PRP-1 affects T-lymphocyte differentiation both in embryonic and postnatal periods of mice thymus development and exerts powerful regulatory effect on myelopoiesis *in vivo* as well. Our experiments showed that during lymphocytopenia in mice caused by cyclophosamide and during P. aeruginosa-caused toxicosis PRP-1 not only abolished lymphocytopenia, but also augmented the content of neutrophils and other cells of myeloid lineage. PRP-1-induced differentiation in the thymus of $TCD4^+$ and $TCD8^+$ lymphocytes during postnatal and embryonic development was shown. It is possible that progenitors of lymphoid and myeloid lines are also involved in the process. Data obtained indicate that PRP-1 probably affects stem cells of myeloid and lymphoid origin. Bone marrow contains two well-characterized types of stem cells: hematopoietic and mesenchimal stem cells. The hematopoietic stem cells normally give rise to the cells of the blood and immune system. The mesenchymal stem cells normally form chondrocytes and osteoblasts[323]. The stem cell is an undifferentiated cell that can divide indefinitely and that produces progeny, which include both more stem cells and cells that are destined to become differentiated ones.

Recently researches' attention was focused on remarkable ability of adult stem cells to produce differentiated cells from embryological unrelated tissues; even fully differentiated cells can change their phenotype (transdifferentiation)[323-333]. For example, hematopoietic progenitor cells that are derived from bone marrow in some conditions could generate astrocytes of CNS[332]. Similar events of transdifferentiation take place under the influence of signal molecules. There is a connection between bone-marrow stem cells and brain cells. Experimental data prove the existence of an old neurohumoral axis, hypothalamus - hypophysis - bone-marrow - thymus, one of mediators of which is PRP-1. Recently we showed the accumulation of PRP-1 in the granulocytes of bone-marrow (unpublished).

The effect of PRP-1 on the differentiation of T-lymphocytes in fetal and neonatal thymus cell cultures, as well as on the regulation of myelopoiesis, is similar to that of transcription factor PU-1, which plays a critical role in the generation of both lymphoid and myeloid lineages. I believe that PRP is a unique transcription factor for production of progenitor cells of lympho- and myelopoiesis or a factor regulating the expression of transcription factors (PU-1, GATA-3, IKAROS, etc.). We can assume that PRP-1 stimulates TH-1 and TH-2 formation (taking into account formation of $TCD4^+$ and $TCD8^+$ lymphocytes) which define the type of immune defense, humoral and cellular immunity. As was shown, PRP-1 is powerful neuroprotector in neurodegenerative pathologies of the brain and spinal cord.

Taking into consideration that all neurodegenerative brain diseases are accompanied to some extent by development of autoimmune diseases, I can conclude that mediators of the neuroendocrine brain immune system (PRP and others) can prevent the development of brain autoimmune pathologies such as encephalomyelitis, spongiform encephalopathy, autoimmune neurological disorders, multiple sclerosis, Alzheimer disease, etc. Cytokine expression in the brain during autoimmune diseases is not precluded and might serve as one of the defense mechanisms. Our main idea was to study the role of brain-borne cytokines in the biochemical mechanisms of neurodegenerative and autoimmune processes in the brain.

H. Besedovsky and coworkers discovered the phenomenon of brain cytokine expression (IL-1, IL-6) during long-term potentiation (LTP). They discovered a reliable increase in IL-1β gene expression, triggered by glutaminergic neurons through NMDA receptors, in hyppocampal slices and in freely-moving rats during the course of LTP. The authors found that the IL-6 gene is also overexpressed during *in vivo* and *in vitro* LTP (Besedovsky et al., 2001)[65].

These data show that cytokines are components of neuronal activity and appear during different functional conditions of neurons and glia. In this respect of importance is the discovery by H. Wekerle and H. Neuman[334] of expression of major histocompitibility complexes I and II during suppression of hippocampus neuron activity by tetradotoxin. Our data show that PRP-1 stimulates expression of interleukins in fibroblasts, astrocytes, and macrophages. Administration of PRP-1 enhanced the expression of interleukins in rat hippocampus and hypothalamus, as well as that of PRP-1 itself. Despite a large body of evidence obtained in our work, there is still not enough information for complete understanding of broad spectrum of PRP molecular action; therefore, it was important to assume the existence of PRP-binding domains in many proteins. In other words, it is possible to accept the suggestion that PRP-1 can produce an effect on many biochemical and immunological systems by direct interaction with PRP-1-binding domain (SH2, SH3, and WW domains)[335-337].

SH2 domain recognize Y-P motif in peptide structure and SH3 homology domains (Src-2 and Src-3) as well as WW domains can recognize proline-rich peptides. In its function the WW domain embodies elements of SH3 and SH2 domains by recognizing proline-rich ligands and being in some cases regulated by phosphorylation. The SH3 domain is the most widespread protein recognition module in the proteome, and more than 1500 different SH3 domains can be identified in protein databases[338,339]. There are eight classes of SH3 domains. They were found in proteins that are implicated in signal transduction, cytoskeleton organization, and membrane traffic[340]. The Src homology-3 (SH3) region is a small protein domain present in a very large group of proteins including cytoskeletal elements and signaling proteins. It is believed that SH3 domains serve as modules that mediate protein-protein association, and along with Src homology 2 (SH2) domains they regulate cytoplasmic signaling[338]. The SH3 domains recognize proline-rich motifs possessing the left-handed type II polyproline (PP II) helix conformation. Two proline residues directly contact the receptor. Other prolines in the ligands appear to function as a molecular scaffold promoting the formation of the PP II helix[339]. Some signaling and transforming proteins may serve as adaptors to link tyrosine kinase to specific target proteins.

These observations suggest that SH2 and SH3 domains participate in the control of intracellular responses to growth factor stimulation[341]. SH2 domains are a common feature of an otherwise diverse group of cytoplasmic polypeptides that serve as intracellular targets of Trks and more complex multi-subunit receptors as those for antigens, cytokines, and extracellular matrix components[336,337]. SH2 domains bind specific phosphotyrosine containing peptide motifs[335-337]. Structural, biochemical, and genetic analysis has revealed that SH2 domains not only recognize phosphotyrosine, but also recognize between three and five residues immediately C-terminal to the phosphotyrosine, in a fashion that varies from one SH2 domain to another. Physiological receptor autophosphorylation (tyrosine phosphorylation) sites can associate with multiple SH2 proteins *in vivo*. SH2 domains are found in a large number of proteins, which otherwise have distinct biochemical activities. 111 SH2 domains are found in proteins of diverse functions, including regulation of protein and lipid phosphorylation, phospholipid metabolism, transcription regulation, cytoskeletal organization, and control of Ras-like GTP-ases[335]. Many functional proteins contains SH2 domains: adaptors, scaffold proteins, kinases, phosphatases, transcription factors, small GTPase signaling molecules, signaling proteins, etc.[335]. SH2 domains bind with phosphotyrosine binding domains (PTB) originally characterized through their ability to recognize phosphorylated Asn-Pro-X-Tyr motifs, such as those found in the RTKs for nerve growth factor[335]. Proteins containing an SH2 domain can inhibit Jak kinases [350].

WW domains (composed of approximately 40 amino acids) also recognize proline-containing ligands and fold as stable triple-strand sheets in the absence of ligands or disulfide bridges. They are found in many different proteins, often localized in the cytoplasm as well as in the nucleus[340,342,343]. WW domains attracted researches' attention because the signaling complexes they mediate have been implicated directly or indirectly in several human diseases including Liddle's syndrome of hypertension, muscular dystrophy, Alzheimer's and Huntington diseases, and more recently, cancer[344]. This is evident for SH2/SH3 adaptors, such as Grb2, Nck, and Crk family members, each of which possesses a single SH2 domain and one to three SH3 domains, and can thereby link a single phosphotyrosine site to multiple intracellular targets with proline-rich SH3 binding sequences[345-347]. "Protein interaction domain families that modulate the formation of macromolecular complexes recognize specific sequences or structural motifs. For instance, SH3 and WW domains bind to polyproline peptides while SH2 and PHA domains bind to peptides phosphorylated in Tyr and Thr, respectively"[344].

Two other proteins that bind PRPs are described in a pathway that regulates CD2-triggered T-lymphocytes (Nishizawa et al., 1998). Recently, a novel family of SH2-containing proteins, SOCS-1, was identified. These proteins are proposed as negative regulators of cytokine signaling. SOCS-1 is primarily expressed in the thymus and plays a critical role in negative selection. In absence of SOCS-1 the development of high number of autoreactive T-cells was observed. SOCS plays a role in inhibiting T-cell activity, which leads to a reduced threshold of T-cell functions, and hence participates in the autoimmune phenotype [348]. In mice, the deletion of SOCS-1 is associated with perinatal lethality so that all the mice die within 2-3 weeks after birth. Thymocyte differentiation is dramatically affected in the absence of SOCS-1. "The possibility that the neonatal pathology associated with the deletion of the SOCS-1 gene was associated with altered thymocyte differentiation, and the generation of antireactive T-cells was examined with two approaches" (Moriggl)[349]. SOCS-1 has a critical role in negative selection during thymic differentiation. It can be used as a marker for cytokine signaling during lymphocyte development. There are 20 proteins containing a C-terminal SOCS box that form five structural classes [351, 352].

The binding of several proteins to the T-cells is mediated by SH2 domain adaptor molecules LAT and SLP-76. They are central players in the processes of T-cell development and activation. For normal T-cell development the existence of leukocyte-specific adaptor protein SLP-76 is very important [353]. Like T-cell activation mechanisms, protein tyrosine kinases (Syk, Btk, and Lyn) play a pivotal role in the B-cell activation mechanisms, as well as tyrosine phosphatases, adaptor proteins SLP-65/BLNK, Grb 2, Cb 1, SH3 P7 and vav, etc. The SLP (SH2 domain-containing leukocyte protein) belongs to adapter family proteins which possess a number of tyrosine phosphorylation sites of the YXXP type[354,355]. Thus, SH2 and SH3 domains play an important role in signaling complexes on T-cell antigen receptor and the LAT-molecule. The localization of LAT in plasma membrane microdomains is required for its tyrosine phosphorylation and consequently, for the binding of critical linkers and enzymes SH2 and SH3 domains, which recognize proline-pich motifs [354].

Analysis of some regulatory biochemical mechanisms of SH2, SH3, and WW protein domains serves as a basis for our understanding the mechanisms of PRP-1 action. PRP-1 generates the respiratory burst in neutrophiles (unpublished data).

The antimicrobial peptide, PR-39 (proline and arginine-rich peptide), is presumed to act by binding to the SH3 domains of the NADPH oxidase subunits, thereby blocking the assembly of a functional enzyme[356]. When neutrophils are incubated with the PR-39 peptide, NADPH oxidase production of superoxide anion O_2^- is blocked.

The T-cell receptor (TCR)-mediated tyrosine phosphorylation and activation of PLCγ-1 leads to hydrolysis of phosphatidylinositol-4,5-bisphosphate (PIP2) to generate lipid second messengers that induce calcium mobilization and activation of the Ras signaling pathway. These events can than trigger the activation of transcription factors such as NF-AT, resulting in the expression of immunologically relevant genes [341]. In T-cell activation adaptor proteins such as LAT (linker for activation of T-cells) and SLP-76 (SH2 - domain - containing leukocyte protein of 76 kDa) play a pivotal role. LAT is an integral membrane protein (36-38 kDa) possessing a cytoplasmic tail sites for SH2-domain-containing proteins[357]. Essential role for LAT and SLP-76 in T-cell activation was demonstrated. To ascertain whether these domains can be functionally important acceptors of PRP, and therefore to explain such a wide spectrum of its influence on biological processes and the activity of many enzymes, is not an easy task and will be the subject of our future studies.

However, the discovery of IL neurosecretion and of a new class cytokines, proline-rich polypeptides, produced by hypothalamic neurosecretory cells NPV and NSO, changed our understanding of the neuroendocrine hypothalamus-hypophyseal system and the role of neurophisin II precursors in the regulation of organism functions. Taking into account that PRP-1 is one of the neurotrophic factors, or neurotrophin-like substances, we can partially explain some of molecular mechnisms of PRP-1 action on the immune and nervous systems.

The expression of NGF receptors by human peripheral blood mononuclear cells as well as by a subpopullation of splenic mononuclear cells was demonstrated[358]. These results would suggest a role for NGF in the regulation of immune function in man, and also in animals. The NGF protein has been shown to be essential for the development of peripheral sympathetic and sensory neurons[280] where it acts on cholinergic magnocellular neurons[359]. The modulation of immune cell reactivity by NGF was demonstrated[360] (mast cells, human neutrophils, etc.). Several reports suggest a role for NGF in the modulation of immune function in man. NGF upregulates the expression of IL-2 receptors on human lymphocytes in the absence of other mitogenic stimuli. Matsuda and coworkers have reported that NGF interacts with T-cells to induce the synthesis of human CSFs[361,362]. NGF has been shown to promote chemotaxis of neutrophils and activated macrophages *in vitro* and *in vivo*[363]. Increased levels of NGF influencing the regeneration of injured neurons are observed in experimental lesions of the nervous system. In *in vitro* studies NGF protects against perturbations of the oxidant-antioxidant balance by the induction of catalase and glutathione transferase[364]. NGF can protect neuronal cells *in vitro* against hydrogen peroxide insults by selective induction of catalase and glutathione transferase. Some authors suggest that the effect of NGF on neuronal viability is mediated by its effect on oxidant-antioxidant balance with particular relevance to neurological diseases.

As was shown by us[242,243], PRP-1 decreases high level of LPO in erythrocytes and lymphocytes and completely normalizes the level of LPO in the heart right ventricle during cardiopulmonary insufficiency. The activity of catalase was inhibited by PRP-1 *in vitro* and *in vivo* in different pathological states. It was reported that NGF binds to chromatin in a displaceable and saturable fashion[365]. The presence of both NGF and NGF receptors has been demonstrated in the brain and SC. Neurotrophins repeatedly modulated neuronal excitability regulating synaptic plasticity in the hippocampus and SC neurons. BDNF was shown to modulate sodium channels in the CA1 region of the hippocampus and enhance synaptic currents in hippocampal postsynaptic neurons[366].

As demonstrated above, PRP-1 is a unique regulator of the activity of caspases which play a significant role in molecular mechanisms of neuronal survival and apoptosis: it causes the inhibition of effector caspases -3 and -9 and activation of caspases -2 and -6.

Discovery in the hypothalamus neurosecretory nuclei (NPV and NSO) of FK-506 immunosuppressor receptors (immunophilins) possessing peptidyl-prolyl cis-trans isomerase activity brings us closer to the understanding of fundamental theoretical questions of the contemporary immunology theory of brain tissue transplantation. If we consider NPV and NSO as main centers of IL biosynthesis in the brain where also immunophilin is formed, we can suggest that FK-506 immunosuppressor (immunophilin ligand), when bound to immunophilin, can inhibit calcineurin (CaN) activity, thus preventing IL-2 biosynthesis. Understanding the role of immunophilins in both biochemistry and physiology of the neuronal membrane will contribute to solving the problem of brain tissue transplantation. As is known, FK-506 and cyclosporin A, when bound to immunophilin and cyclophilin, inhibite calcineurin activity, which results in the inhibition of biosynthesis of IL-2 in T-lymphocytes. IL-2 gene transcription is regulated by the nuclear factor of activated T-cells (NF-AT). Cytoplasmic form of NF-AT is phosphorylated by CaN and must be dephosphorylated in order to be translocated to the nucleus and activate the IL-2 gene transcription. As was mentioned in the introduction, the addition of 10 - 100 pM of FK-506 to a cell culture treated with 50 mM beta-amyloid peptide (25 - 35) resulted in about 70% protection against beta-amyloid neurotoxicity of hippocampal neurons.

These data allow us to draw a parallel between lymphocytes and neuronal cells (NSO, NPV, and hippocampus), regarding similar mechanisms of IL formation and inhibition. If the hypothesis that NSO and NPV cells execute lymphocyte function proves to be true, this may also be an argument in favor of a neuroendocrine immune system of the brain. As was convincingly demonstrated, PRP-1 affects the expression of a number of ILs, lymphocytes, astrocytes, fibroblasts, and GH in BALB/c cells, fibroblasts, and even in CNS neurons (for instance, PRP-1 injection into rats induces its processing in the brain).

The aim of this book is to solve the problem of whether the brain itself is an immune organ, and also to define cellular, neurochemical, and immunological properties of the brain neuroendocrine immune system which is used for immune defense during different infections and pathological states.

The discovery of the brain neuroendocrine immune system (biosynthesis of signal molecules in NPV and NSO immunocompetent cells, elucidation of their primary structure, and their chemical synthesis) turns a new page in our understanding of molecular mechanisms of the immune system regulation by neurosecretory cytokines and ensuring neuronal survival.

SUMMARY

The discoveries of interleukin neurosecretion, of novel cytokines including PRPs, immusuppressor macrolide FK-506 receptor (immunophilin), thymosin β_1 produced by magno- and parvocellular cells of NPV and NSO, macrophage migration inhibitory factor (one of the first cytokines found to be secreted from lymphocytes), thymosin β_4 (1-39), parvalbumin alpha in the hypothalamus, and the involvement of all above factors in the regulation of immune system confirm significant role of signal molecules of brain neuroendocrine immune system in molecular mechanisms of immune response, neuronal survival, HPA-axis, and adaptation[31-34,212].

PRP-1 is the regulator of both humoral and cellular immunity, differentiation of thymocytes, and myelopoieses. It possesses strong antiviral (*in vitro*) and antibacterial activity (*in vivo,* against *Salmonella typhimurium*, *S. cholerae suis*, *S. typhi*, *E. coli*, *P. aeruginiosa*, *Shigella Flexneri*, *Shigella Sonnei*, *St. aureus*, *Streptococcus pneumoniae*, etc.). PRP-1 can be recommended in future for treatment of immunodeficiencies and different infectious diseases. It protects against aluminum neurotoxicity and abolishes the accumulation of this element in brain microstructures and visceral organs (electron-microscopic studies). Data obtained show that PRP-1 can be also recommended for treatment and prophylaxis of aluminosis. PRP-1 expresses antineurodegenerative activity in the cases of SC hemisection, trauma, crush syndrome as well as in aluminum neurotoxicosis. PRP-1 also possesses antitumor activity *in vitro* against Jurkat cells, L_{929} fibroblasts, neurinoma Hasser's ganglion cells (this study), and *in vivo* against sarcoma-45 (C45) (unpublished data).

PRP-1 metabolic, neuroprotective, anti-infectious action can explain its anti-oxidative – anti-free-radical, neurotrophic – anti-neurodegenerative, immunotrop - antibacterial and antiviral properties.

REFERENCES

1. Scharrer E., and Scharret B. 1954. Recent Progress in Hormone Research 10:183-190.
2. Vigneaud V., Gish D.T., and Kotsoyanis P.G. 1954. J. Amer. Chem. Soc. 76:47-51.
3. Guillemin R. 1977. Pages 1-28, *in*: Roy O., Gree P. (eds.), Recent Progress in Hormone Research Academic Press, 33(4).
4. Schally A.V., Coy D.H., Meyers Ch.A. 1978. Annu. Rev. Biochem. 47:1-81.
5. Schally A.V., Cumaru-Schally A.M., Nagy A., Kovacs M., Szepeshazi K., Plonowski A., Varga J.L., and Halmos G. 2001. Frontiers in Neuroendocrinology. 22:248-291.
6. Galoyan A.A. 1962. Reports Acad. Nauk Arm. SSR. 34(4):109-111.
7. Galoyan A.A. 1965. Some Problems of Biochemistry of Hypothalamic Regulation (in Rus.), Hayastan,Yerevan. 236p.
8. Galoyan A.A., Rostomian M.A. 1967. Biol. J. Armenii 20(9):3-7.
9. Srapionyan R.M. 2001. Pages 56-60, *in* Galoyan A. (ed.), Proceedings of the International Conference. "Biochemical and Molecular Biological Aspects of Immune System", Encyclopedia Armenica Publishing House, Yerevan.
10. Galoyan A.A. 1964. The regulation of Neurosecretion and Hormones of Hypothalamo-Neurohypophyseal system. Doctor of Scince Thesis.
11. Galoyan A.A. 1964. Dokl. Acad. Nauk ArmSSR. 38(5):305-308.
12. Galoyan A.A. 1975. Pages 651-657, *in* Walter R., Meienhofer J. (eds.), Proceeding of the Fourth American peptide symposium. Peptides. Chemistry, Structure, Biology, Ann Arbor Science publishers.
13. Galoyan A.A., Srapionian R.M., Sahakian F.M. 1978. Pages 190-193, *in* Bargman W., Oksche A., Polenov A., Scharrer B. (eds.), Neurosecretion and Neuroendocrine Activity. Evolution, Structure and Function. Springer-Verlag, Heidelberg
14. Galoyan A.A. 1978. Ergebnisse der Experimentallen Medizin, 29:99-115.
15. Galoyan A.A., Gurvits B.Ya., Saribekian G.A., Kirakosova A.S. 1979. Pages 165-181, *in* Cehovic G., Robinson G. (eds.), Cyclic Nucleotides and therapeutic Perspectives, Pergamon Press.
16. Srapionian R.M., Sahakian F.M., and Galoyan A.A. 1981. Neurochem. Res., 6(12):1299-1308.
17. Galoyan A.A., Srapionian R.M. 1982. *In* Chemistry of Peptides and Proteins. Walter de Gruyter, 10(4):103-107.
18. Galoyan A.A., and Srapionian R.M. 1983. Neurochem. Res. 8(12):1511-1535.
19. Abrahamyan S.S., Fodor M., Galoyan A.A., and Polkovits M. 2002. Neurochem. Res. 27(5):380-388.
20. Galoyan A.A. 1986. Neurochem. Res. 11(6):769-785.
21. Srapionian R.M., Mardanian S.S., Sahakian S.A., Karapetian R.O., Sahakian F.M., and Galoyan A.A. 1987. Biochemistry (USSR), 52(1):89-95.
22. Galoyan A.A., Kevorkian G.A., Voskanian L.O., Alexanian S., Muradian M. 1988. Neurochem. Res., (USA) 13(5):493-498.
23. Galoyan A.A., Gurvits B.Ya., Sharova N.P. 1989. Neurochem. Res. 14(12):1213-1221.
24. Srapionian R.M., Gabrielian S.K., Abelian J.G., and Galoyan A.A. 1991. Exper. Clin Med. 31(1):672-677.
25. Galoyan A.A. 1992. News in Physiological Sciences, 12(7):279-283.
26. Galoyan A.A., Srapionian R.M. 2001. Neurochemistry (RAS & NAS RA). 18(3):206-211.
27. Galoyan A.A., Srapionian R.M., Arora R.C., and Armour J. 2001. Autonomic Neuroscience: Basic and Clinical. 92:11-20.
28. Galoyan A.A. 1997. Biopolymers Peptide Science. Ed by V.T. Ivanov. John Wiley and Sons, 43:135-139.
29. Srapionian R.M., Sahakian S.A., Karapetian R.O. Sahakian F.M., Markin V.A., Odabashian A.B., Galoyan A.A. 1988. Neurochemistry (USSR) 7(3):436-439.
30. Galoyan A.A. 1968. Vestnik Akad. Nauk USSR 12(5):67-73.
31. Galoyan A.A. 1997. Biochemistry of Novel Cardioactive Hormones and Immunomodulators of the Functional System Neurosecretory Hypothalamus - Endocrine Heart. Nauka Publishers (in English), Moscow, 240p.
32. Galoyan A.A. 2000. Neurochem. Res. 25(9/10):1343-1355.
33. Galoyan A.A., Aprikyan V.S. 2002. Neurochem. Res. 27(4):305-312.
34. Aprikyan V.S., and Galoyan A.A. 2001. Pages 183-189, *in* Galoyan A. (ed.), Proceedings of the International Conference. "Biochemical and Molecular Biological Aspects of Immune System", Encyclopedia Armenica Publishing House, Yerevan.

35. Galoyan A.A., Gurvits B.Ya. (1992), Neurochemistry (USSR5), 11,2,8991.
36. Galoyan A.A. 2001. Pages 22-34, *in* Galoyan A. (ed.), Proceedings of the International Conference. "Biochemical and Molecular Biological Aspects of Immune System", Encyclopedia Armenica Publishing House, Yerevan.
37. Galoyan A.A., Gurvits B.Ya., Shuvalova L.A., Davis M.Y., Shively J.E., Lee T.D. 1992. Neurochem. Res. 17(8):773-777.
38. Brightman, M.W., Ishihara S., and Chang L. 1995. Curr. Top. Microbiol. Immunol. 202:63-78.
39. Kreutzberg G.W. 1996. Trends Neurosci 19:312-3118.
40. Laye S., Bluthe R.M., Kent S., Lombe C., Medina C., Parnet P., Kelley K., and Dantrer R. 1995. Am. J. Physiol. 268 (Pt2):R1237-R1331.
41. Saphir D., Abramsky O., Mor G., and Ovadia 1987. Brain Behav. Immunol. 1:40-51.
42. Besedovsky, H.O., and del Rey, A. 1996. Endocrine Rev., 17:64-102.
43. Dunn A.A., Powell M.L., Meitin C., and Small Jr., P.A. 1989. Physiol. 45:591-594.
44. Fleshner M., Goehler L.E., Hermann J., Relton J.K., Maier S.F., and Watkin L.R. 1995. Brain Res. Bul. 375:605-610.
45. Markossian K.A., Gurvits B.Ya., Galoyan A.A. 1999. Neurochemnistry (RAS & NAS RA) 16(1):22-25.
46. Ohnishi S.T., Barr, J.K. 1978. Anal. Biochem. 86:193-200.
47. Kaye J.,Gillis S., Mizel S.B. et., 1984. J. Immunol. 133:1339-1345.
48. Kaye J., Porcelli S., Tite J., et al. 1983. J. Exp. Med. 158:836-840.
49. Kurt-Jones E.A., Beller D., Mirel S.B. 1985. Proc. Natl. Acad. Sci. USA. 82:1204-1208.
50. Simon P.L., Laydan J.T., Lee J.C. 1985. Immunol. Methods. 84:85-94.
51. Serrate S.A. et al. 1987. J. Immunol. 139:2338-2343.
52. Aarden L.A., De Groot E.A., Schaap O.L. et al. 1987. Eur. J. Immunol. 17:1411-1415.
53. Hoffman P., Bessler W.G. 1989. Biol. Chem. Hoppe-Seyler. 370:575-579.
54. Sugarman B.J., Aggarwab B.B., Haas P.E. et al. 1985. Sciences 230:943-944.
55. Turnbull A.V., Pitossi F., Lebrun, J., Lee S., Meltzer J., Nancli D. del Rey A., Besedovsky H.O., and Rivier C. 1997. J. Neuroscience. 117:3262-3273.
56. Aloisi F., Penna G., Cerase J. 1997. J. Immunol. 159(4):1604-1613.
57. Virgin H.W., Kurt-Jones E.A., Wittenberg G.E. et al. 1985. J. Immunol. 135:3744-3749.
58. Johnoson H.M., Farrar W.L., Torres B.A. 1982. J. Immunol. 129:983-986.
59. Plata-Salaman C.R. 1989. Brain Behav. Immun. 3(3):193-213.
60. Weigent D.A., Blalock J.E. 1985. Leukoc. Biol. 58:137-150.
61. De Simoni M.G., Imeri L. 1998. Biol. Signals Recept. 7:33-44.
62. Del Rey A., Besedovsky H.O. 1992. Eur. J. Clin. Invest. 22:10-15.
63. Berczi I., Chalmers I.M., Nagy E. et al. 1996. Daillieres Clin. Rheumatol. 10:227-257.
64. Schedlowski M., Schmidt R.E. 1996. Naturwissenschaften. 83:214-220.
65. Besedowsky H.O., Balchun D., Pitossi F., Schneider H., and del Rey A. 2001. Pages 41-46, *in* Galoyan A. (ed.), Proceedings of the International Conference. "Biochemical and Molecular Biological Aspects of Immune System", Encyclopedia Armenica Publishing House, Yerevan.
66. Rothwell N.J. 1999. J. Physiol. 514(1):3-17.
67. Turnbull A.V., Rivier C. 1995. Brain. Behav. Immun. 9:253-275.
68. Rivier C. 1995. Front. Neuroendocrinol. 16:151-182.
69. Lee Y.B., Kim S.U. 1997. J. Neurochem. 69: S178.
70. Galoyan A.A., Sahakian F.M. 1971. Dokladi Acad. Nauk USSR 201(2):483-485.
71. LaBella F.S. 1968. Canad. J. Physiol. Rharmad. 46, 2, p. 335-345.
72. Richter D. 1985. Neurobiology of Vasopressin. Current Topics in Neuroendocrinology. Ganten D., Pfaff D. (eds), 4:1-16.
73. Watson S.J., Seidah N.G., Chretien M. 1982. Science 217:853-855.
74. Lauber M., Canier M., and Cohen P. 1979. FEBS Letters. 97(2):343-347.
75. Nicolas P. et al. 1981. J. Biol. Chem. 256(17):9289-9294.
76. Seidah N.G., Benjannet S., Chritien M. 1981. Biochem. Biophys. Res. Commun. 100(2):901-907.
77. Land H., Schüz G., Schmale H., and Richter D. 1982. Nature 295:299-303.
78. Holwerda D.A. 1972. Eur. J. Biochem. 28:340-346.
79. McCann S.M., Karanth S., Kamat A., et al. 1994. Neuroimmunomodulation. 1:2-13.
80. Smith D.G., and Massay D.E. 1979. Biochem. Biophys. Res. Commun. 87(4):1006-1010.
81. Nagy G., Mulchahey J., Smith D.G., and Neil J.D. 1988. Biochem. Biophys. Res. Commun. 151(1):524-529.
82. Burbach J.P., Seidah N.G., and Chretien M. 1986. Eur. J. Biochem. 156:137-142.
83. Koff W.C., Dunegan M.A. 1986. J. Immunol. 135:150-154.

84. Tager H.S., Emdin S.O., Clark J.L., and Steiner D.F. 1983. J. Biol. Chem. 248:3476-3482.
85. Seidah N.G., Lazure C., Chrétien M., Thibault G., Garsia R., Cantin R., Genest G., Palaveda A., Colton W.J., Nutt S.R., Brady S.F., Lyle T., Ciccarone T.M., and Veber D.F. 1984. Proc. Natl Acad. Sci. USA. 81:2640-2644.
86. Austen B.M., Smyth D.G., and Snell C.R. 1977. Nature. 269:619-621.
87. Burbach J.P.H., Loeber J.G., Verhoef J., Wiegant V.M., de Kloet E.R., and de Vied D. 1980. Nature (Lond.) 283:96-97.
88. Peach M.J. 1977. Physiol. Rev. 57:313-370.
89. Birch N.O. et al. 1994. Neuropeptides. 27:307-322.
90. Coates L.C., Birch N.P. 1998. J. Neurochem. 70:1670-1678.
91. Azaryan A., Galoyan A.A. 1987. Neurochem. Res. 12(2):207-213.
92. Burbach P.H. 1984. Pharmacol. Ther. 24:321-354.
93. Pickering B.T., Jones C.W., Burford G.D. McPherson M., Swann R.W., Heap P.E., and Morris J.F. 1975. Neurophysins: Carriers of peptide hormones. Ed. by R. Walter. Annals of the New York Academy of Sciences, 248:15-35.
94. McKelvy J.F. 1975. Annals of the New York Academy of Sciences. 248:80-91.
95. Poinser A.M., and Douglus W.W. 1968. Mol. Pharmacol. 4:531.
96. Douglus W.W. , Poinser A.M. (1965) Nature,208,5015,1102-1103.
97. Norstom A.H. et al. 1971. Z. Zellforsch. 113:271.
98. Murai J., and Bon-Jonathan N. 1987. Endocrinology. 121:205-211.
99. Nilavar G. et al. 1980. Neuroendocrinology. 30:15-158.
100. Graay T.S. et al. 1987. J. Comp. Neural. 262:365-374.
101. Reinharz A.A., and Vallotton M.B. 1977. Endocrinology. 100(4):944-100111.
102. White W.F. et al. 1974. Endocrinology. 94:14-22.
103. Abrahamyan S.S., Meliksetyan I.B., and Galoyan A.A. 2001. Pages 62-73, , *in* Galoyan A. (ed.), Proceedings of the International Conference. "Biochemical and Molecular Biological Aspects of Immune System", Encyclopedia Armenica Publishing House, Yerevan.
104. Palkovits M., Zaborsky L., Feminger A., Mezey E., Fetecte M.I.K., Herman J.P., Kanyicka B., and Szabo D. 1980. Brain Res. 191:161-171.
105. Swanson I.W., Sawchenko P.E., Berod A., Hartman B.K., Helle K.B., and Van Orden D.E. 1981. J. Comp. Neurol. 196:271-285.
106. Lecham R.M., Nestler J.L., and Jacobson S. 1982. Brain Res. V. 24 5, pp. 7-15.
107. Wregand S.J., Price J.L. 1980. J. Comp. Neurol. V. 192, pp. 1-19.
108. Palkovits M. 1984. *In* Muller E.E., and MacLeod R.M. (eds.), Neuroendocrine perspectives, Elsevier, Amsterdam, 3:1-69.
109. Galoyan A.A., Bedian V., Melikyan N., Vartazaryan N. 2001. Neurochemistry (RAS & NAS RA). 18(2):147-152.
110. Bedian V., Oliver C.E., Mc Coon P., Kaufman S.A., 1986. Dev. Biol. V. 115, pp. 105-118.
111. Aghajanov M.J., Vahgradyan H.G., Zilfyan A.V., Avakyan S.A., and Galoyan A.A. 2001. Pages 82-85, *in* Galoyan A. (ed.), Proceedings of the International Conference. "Biochemical and Molecular Biological Aspects of Immune System", Encyclopedia Armenica Publishing House, Yerevan.
112. Snow E.C., 1985. J. Immunol. 135(Suppl.2):776.
113. Strom J.B., Bangs J.D. 1982. J. Immunol. 128:155.
114. Sveshnikov P.G., Gromova N.V., Rubtsov P.M. et al. 1988. Human Growth Hormone. Pushchino Sci. Centre Biol. Invest. of USSR, p. 60 (in Rus.).
115. Aprikyan V.S., Galoyan K.A., and Galoyan A.A. 1999. J. Neurochem. 77(Suppl.): S68D.
116. Aprikyan V.S., Galoyan A.A. 1999. Medical Science of Armenia. 39(2):23-30.
117. Aprikyan V.S., Galoyan A.A. 1999. Medical Science of Armenia. 39(4):29-36.
118. Aprikyan V.S., Galoyan A.A. 2000. Neurochemistry (RAS & NAS RA) 17:60-63.
119. Aprikyan V.S., Galoyan A.A. 1999. Report of Armenian National Academy of Sci. 99(4):367-371.
120. Finney, D.J. 1952. Probit analysis. A statistical treatment of the sigmoid response curve. 2 ed. Cambridge Univ. Press, Cambridge.
121. Boman, H.G., Agerbath, B., and Boman A. 1993. Infect. Immun. 61:2978-2984.
122. Boman, H.G., 1995. Annu. Rev. Immunol. 13:61-92.
123. Boman H.G., Brockaet W.F., 1998. The immunologist, 6(6):234-238.
124. Hancock R.E.W., and Lehrer R. 1998. Trends Biotechnol. 16:82-88.
125. Ahmed I.W., Perkins R., Lupan D.M., Selsted M.E., and Janoff A.S. 1995. Biochem. Biophys. Acta, 1237:109-114.

126. Aley S.B., Zimmerman M., Hetsko M., Selsted M.E., and Gillin F.D. 1994. Infect. Immunol. 62:5397-6403.
127. Schluesner H.J., Badermacher S., Melms, A., and Jung J. 1993. J. Neuroimmunol, 47:199-202.
128. Fallf T.J., Karunaratne, D.N., Harcock R.E.W. 1996. J. Biol. Chem. 271, 19298-19303.
129. Boman, H.G. 1998. Scand. J. Immunol. 48(1):15-25
130. Suh, J.-Y., Lee, K.-H., Chi, S.-W., Hong S-Yu, Choi B.-W., Moon H.-M., Choi B.-S. 1996. FEBS Letters, 392:309-312.
131. Park, C.B., Kim, M.S., and Kim, S.C., 1998. FEBS Letters, 507:95-100.
132. Park C.B., Kim, M.S., and Kim S.C. 1996. Biochem. Biophys. Res. Commun. 218:408-413.
133. Kobayashi, S., Takasxhima K., Park Ch.B., Kim S.B., and Matsuzaki K., 2000. Biochemistry, 39(29):8648-8654.
134. Kim H.C., Park C.B., Kim M.S.W., and Kim S.C. 1996. BBRC, 229:381-387.
135. Steiner H. 1982. FEBS Letters, 137:283-287.
136. Alexander M., Kim Y.-H., Tahk S., Hong T., Weis P., Waring A.J., Ganz T. 2001. FEBS Letters, 504:5-10.
137. Galoyan A.A., Kamalyan L.A., Gasparian M.G. 2000. Reports of Arm. Natl. Acad. Sci. 100(3):276-282.
138. Khamblin A., O'Harra A. 1990. Lymphocyti Metodi, pp. 310-338.
139. Davtyan T.K., Blinova A.J., Ignatova T.N., Alexanyan Yu., Meliksetyan M.B. 1993. Bull. Exp. Med. Biol. (in Rus). 116(12):616-618.
140. Davtyan T.K., Alexanyan Yu.T., Grinchuk T.M. 1994. Immunologiya. (in Rus.) 5:33-37.
141. Galoyan A.A., Schakhlamov V.A., Bogdanova J.M., Malaitzev V.V., Michalyova L.M. 2002. Neurochemistry (issue of RAS & NAS RA), 19(1):41-45.
142. Gilliland L.K., Schieven G.L., Noris N.A. et al. 1992. J. Biol. Chem. 267:13610-13615.
143. Sieh M., Batzer A., Schlessinger J., Weiss A. 1994. Mol. Cell. Biol. 14:4435-4439.
144. Fukazawa T., Reedquist K.A., Panshamoorthy G. et al. 1995. J. Biol. Chem. 270:20177-20184.
145. Goldsmith M.A., Dazin P.E., Weiss A. 1988. Proc. Natl. Acad. Sci. (USA), 8:8613-8617.
146. Sherr C.J., Roberts J.M. 1995. Genes and development. 9:11449-1153.
147. Samuelson L.E., Bunnelle S.C., Trible R.P.,Yamazaki T., and Zhang W. 1999. Gold Spring Harbor Laboratory Press, 69:259-263.
148. Robinson J.H., and Jordan R.K. 1983. Immunol. Today. 4:41-45.
149. Robey E., and Fowlkers B.L. 1994. Annu. Rev. Immunol. 12:675-705.
150. Von Bochmer H., and Fehling H.J. 1997. Annu. Rev. Immunol. 12:433-452.
151. Kaufman M.H. 1992. The Atlas of mouse Development Acad Press. San Diego, USA.
152. Ceredic R. 1988. J. Immunol. 141:355-362.
153. Sugawara T., Di Bartolo V., Miyazaki T., Nakauchi H., Acuto O., and Takahama Y. 1998. J. Immunol. 161:2888-2894.
154. Tkahama Y., Hasagawa T., Itohara S., Ball E.L., Sheard M.A., and Hashimoto Y. 1994. Int. Immunol. 6:1505-1514.
155. Tarahama Y., and Nakauchi H. 1996. J. Immunol. 157:1508-1513.
156. Leo O. et al., 1986. J. Immunol. 137:3874-3880.
157. Dazynkiewsicz Z., Robinson J.P., and Crissman H.A. (eds.). 1994. Flow cytometry 2nd ed. Academic press. San Diego,USA.
158. Fowlker B.J., and Pardoll D.M. 1989. Adv. Immunol. 44:207-265.
159. Weismann I.L. 1994. Cell. 76:207-218.
160. Kruisbeek A.M., and Storb U. (eds.). 1996. Lymphocyte development. Curr. Opin. Immunol. 8:155-1254.
161. Durum S.K., and Muegge K. (eds.). 1997. Cytokine Knockout. Humana Press, Tutowa, USA.
162. He Y.W., and Malek T. 1996. J. Exp. Med. 184:289-293.
163. Theze J. (ed.). 1999. The Cytokine Network and Immune Functions. Oxford Univ. Press Inc. New York, USA, 373 p.
164. Kisielow P., and von Boehmer H. 1991. Adv. Immunol. 58:87209.
165. Rodewald H.-R., Kretzschmar K., Swat W., and Takeda S. 1995. Immunity. 3:313-319.
166. Akashi K., Kondo M., Cheshier S., Shizuru J., Gandy K., Domen J., Mebius R., Traver D., and Weissman I.L. 1999. Cold Spring Harbor Symposia on Quantitative Biology. LXIV:1-12.
167. Kondo M., Wessman I.L., and Akashi K. 1997. Cell 91:661.
168. Singh H., DeKoter R.P., and Walsh J.S. 1999. Cold Spring Harbor Symposia on Quantitative Biology. LXIV:13-20.
169. Robey E. et al., 1994. Cell 87:183.

170. Pear W., Tang Z., De Rocco S., Allman D., Hardy R., Pui J., Xu L., Karnell F., Aster J., and Kadesch T. 1999. Cold Spring Harbor Symposia on Quantitative Biology. LXIV:33-38.
171. Flavell R.A. 1999. Cold Spring Harbor Symposia on Quantitative Biology. LXIV:603-610.
172. Greaves M., Chan L., Furley A., Watts S., and Molgaard H. 1986. Blood. 67:1-11.
173. Lieschke G.J., Crail D., Hodgson G., Metcalf D., Stanley E., Cheers C., Fowler K., Basu S., Zhan Y.F., and Dunn A.R. 1994. Blood. 84:1737-1746.
174. Cronkite E.P. 1979. Clin. Hematol. 8:315-370.
175. Davis R.E., Longacre T.A., and Cornbleet P.J. 1994. Am. J. Clin. Pathol. 102:202-211.
176. Pitrak D.L. 1997. Curr. Opin. Hematol. 4:183-190.
177. Finlay B.B., and Falkow S. 1997. Microbiol. Mol. Biol. Rev. 61:136-139.
178. Klainer A.S. 1981. P. 13-28. In: J.P. Sanford and J.P. Luby (eds.) Infection disease. Grune and Stratton Inc., New York.
179. Estey E. 1998. Curr. Opin. Oncol. 10(1):23-30.
180. Fixe P., and Praloran, V. 1997. Eur. Cytokine Netw. 8:125-136.
181. Holden I.A., and Haidaris C.G. 1979. Can. J. Microbiol. 24:593-599.
182. Leijh P.C.J., van Furth R., and van Zwet T.L. 1986. *In* D.M. Weir. (ed.), Handbook of experimental Immunology. 2. Cellular Immunology, edited by D.M. Weir. 4 ed, Blackwell Sci. Publ., Oxford. 46:1-21.
183. Spyridominis A., Mertelsmann R., and Henschler R. 1996. Curr. Opin. Haematol. 3:321-328.
184. Welte T., Leitenberg D., Dittel B.N. et al. 1999. Cold Spring Harbor Symposia on Quantitative Biology. 64:291-302.
185. Shimoda K., Feng J., Muzakami H., Nagata Sc., Watling D., Rogers N.C., Stark G.R., Kerr I.M., and Ihle J.N. 1997. Blood 90(2):597-604.
186. Zhu X., Yeadon J.E., and Burden S.J. 1994. Mol. Cell. Biol. 14:8051-8057.
187. Scott E.W., Fisher R.C., Olson M.C., Kehrli E.W., Simon M.C., and Singh H. 1997. Immunology. 6:437-441.
188. De Koter R.P., Walsh J.C., and Singh H. 1988. EMBO J. 17:4454.
189. Verfaillie C.M., Benis A., Lida J. et al. 1994. Blood. 84:1802-1811.
190. Nichogiannopoulou A., Trevisan M., Neben S., Fredrich C., and Georgopoulos K. 1999. J. Exp. Med. 199:1201.
191. Schatz D.G., Oettinger M.A.and Baltimore D. 1989. Cell. 59:1035.
192. Koipally J., Kim J., Jones A. et al. 1999. Cold Spring Harbor Symposia on Quantitative Biology. 64:79-86.
193. Schneider E., and Dy M. 1999. Pages 146-161, *in* Thezse J. (ed.), The Cytokine Network and Immune Functions, Oxford Univ. Press, Inc., New York.
194. Boulton T.G., Stahl N., and Yancopoulos G.D. 1994. J. Biol. Chem. 269:11648-11655.
195. Pennica D., Wood W.J., and Chien K.R. 1996. Cytokine Growth Factor Rev. 7:81-91.
196. Orkin S.H. 1996. Curr. Opin. Genetics and Development 6:597-602.
197. Strelkov L.A., Mikhailova A.A., Fonina L.A., Petrov R.V. 2000. FEBS Lett. 470:281-284.
198. Karlsson E. 1979. Pages 159-212, chapter 5, *in* Chen-Yuan (ed.), Chemistry of Protein Toxins in Snake Venoms Springer-Verlag, Berlin-Heidelberg-New York.
199. Slotta K. 1955. Prog. Chem. Org. Nat. Prod. 12:406-465.
200. Kaiser E., and Michel H. 1958. Die Biochemie der tierischen Gifte, Franz. Deuticke, Wien.
201. Suzuki T., and Iwanaga S. 1970. *In* Erdos E.G. (ed.), Handbook of Experimental Pharmacology, Springer-Verlag, Berlin-Heidelberg-New York,. 25:193-212
202. Meaume, J. 1996. Toxicol. 4:25-58.
203. Minton S.A. 1974. *In* Thomas, Ch.C. (ed.) Venom Diseases, Springfield (III).
204. Stenlake J.B. 1986. Pages 263-276, *in* Kharkevich D.A. (ed.), Handbook of Experimental Pharmacology Springer-Verlag, Berlin.
205. Hughes R. 1986. Pages 249-253, *in* Kharkevich D.A. (ed.), Handbook of Experimental Pharmacology Springer-Verlag, Berlin.
206. Bowman W.C., and Sutherland G.A. 1986. Pages 419-443, *in* Kharkevich D.A. (ed.), Handbook of Experimental Pharmacology Springer-Verlag, Berlin.
207. Jankovic J., and Hallett M. 1994. Therapy and botulinum toxin, New York: Marcel Decker.
208. Cushman D.W. et al. 1980. Pages 2310-2478, *in* Sandler M. (ed.), Enzyme inhibitors and drugs, Macmillan, London.
209. Rudy, B. 1988. Neuroscience. 225:729-749.
210. Galoyan A.A., Sarkissian J.S., Kipriyan T.K., Sarkissian E.J., Grigoryan Y.Kh., Sulkhanyan R.M., and Khachatrian T.S. 2000. Neurochemical Research. 25(12):1567-1578.

211. Galoyan, A.A., Kipriyan, T.K., Sarkissian, J.S., Sarkissian, E.J., Grigorian, Y.Kh., Andreasian, A.S., and Chavushian, V.A. 2000. Neurochemical Research. 25(6):791-800.
212. Galoyan A.A., Sarkissian J.S., Kipriyan T.K., Sarkissian E.J., Chavushyan E.A., Sulkhanyan R.M., Meliksetyan I.B., Abrahamyan S.S., Grigorian Y.Kh., Avetisyan Z.A., Otieva N.A. 2001. Neurochem. Res. 26(8):1023-1038.
213. Sarkissian J.S., Kipriyan T.K., Sulkhanyan R.M., and Galoyan A.A. 2001. Pages 95-108, , *in* Galoyan A. (ed.), Proceedings of the International Conference. "Biochemical and Molecular Biological Aspects of Immune System", Encyclopedia Armenica Publishing House, Yerevan.
214. Sarkissian J.S., Kipriyan T.K., Grigorian Y.Kh., Sarkissian E.J., Amiryan S.V., Chavushyan E.A., Avetirsyan Z.A. 1999. Vestnik IAELPS. St.-Petersburg (Yerevan) 15:127-132.
215. Macaya A., Munell, F., Ferrer I., de Torres C., and Reventos J. 1988. Molecular Brain Research. 56(1-2):29-37.
216. Abrahamyan S.S., Meliksetyan I.B., Sulkhanyan R.M., Sarkissian J.S, and Galoyan A.A. 2001. Neurochem. Res. 26(11):1225-1230.
217. Kato, N., Halprin, K.M., Matsuo, S., and Taylor, J.R. 1985. A. Biochem. Biophys. Res. Commun. 130:(2)761-767.
218. Sharma H.S., Nyberg F. Gordh T., Alm. p., and Westman J. 2000. Amino acids. 19:351-361.
219. Bregman B.S., McAtee M., Dai H.N., and Kuhn P.L. 1997. Exp. Neurol. 148:475-494.
220. Weidner N., Blesch, A., Grill R.J., and Tuszynski M.H. 1999. J. Comp. Neurol. 413:495-506.
221. Corse A.M., Bilak M.M., Bilak S.R., Lehar, M., Rothstein J.D., and Kuncl, R.W. 1999. Neurobiol. Dis. 6:5.
222. Jiang Y., McLennan I.S., Koishi, K., and Hendry, I.A. 2000. Neuroscience 97:735-423.
223. Friedman B., Kleinfeld D., Ip, N.Y., Verge V.M., Moulton R., Boland P., Zlotchenko E., Lindsay R.M., and Liu L. 1995. J. Neurosci. 15:1044-1056.
224. Diener P.S., and Bregman B.S. 1994. Neuroreport 5:1913-1917.
225. Schmalbruch H., and Rosenthal A. 1995. Brain. Res. 700:254-260.
226. Novikova L., Novikov L., and Kellerth J.O. 1996. Neurosci. Lett. 220:203-206.
227. Neff N.T., Prevette D., Houenou L.J., Lewis M.E., Glicksman M.A., Yin Q.W., and Oppenheim R.W. 1993. J. Neurobiol. 24:1578-1588.
228. Iwasaki Y., and Ikeda K. 1999. J. Neurol. Sci. 169:148-155.
229. Pu S.F., Zhang H.X., Marsh D.J., and Ishi D.N. 1999. J. Neurosci. Res. 55:9-16.
230. Houle S.D., and Ye S.H. 1999. Neuroscience. 94:929-936.
231. Tsitouras P.D., Zhong Y.G., Spungen A.M., and Bauman W.A. 1995. Horm. Metabol. Res. 27:287-292.
232. Li X.B., Chen, M.J., Lei D.Q., Yang B., Lia O.G.S., Shu Y.Y., and Tang S.H. 1999. J. Nat. Toxins 8:359-362.
233. Fletcher J.E., and Middlebrook J.L. 1986. Toxins 24:91-99.
234. Faure G. 2000. Biochime. 82:833-840.
235. Kolko M., DeCoster M.A., de Turco E.B., and Bazan N.G. 1996. J. Biol. Chem. 271:32722-32728.
236. Nigam S., and Schewe T. 2000. Biochim. Biophys. Acta. 1988:167-181.
237. Taoka Y., and Okajima K. 1998. Spinal cord injury in the rat. Progress in Neurobiology. 56:341-358.
238. Balu A.S., and Gowda T.V. 1995. Inflamm. Res. 44:340-344.
239. Costa L.A., Fornari M.C., Berardi V.E., Miles H.A., and Diez R.A. 2000. Immunol. Lett. 75:137-141.
240. Krizaj J., Dolly J.O., and Gubensek F. 1994. Biochemistry. 33:13938-13945.
241. Urushitani M., Inoue R., Nakamizo T., Sawada H., Shibasaki H., and Shimohama S. 2000. J. Neuroci. Res. 61:443-448.
242. Galoyan A.A., Ghazaryan P.A., Ghazaryan A.P. 2001. Neurochemistry (RAS & NAS RA). 18(4):379-386.
243. Ghazaryan P.A., Ghazaryan A.P., and Galayan A.A. 2001. Pages 115-126, *in* Galoyan A. (ed.), Proceedings of the International Conference. "Biochemical and Molecular Biological Aspects of Immune System", Encyclopedia Armenica Publishing House, Yerevan.
244. Chen R., and Robinson S.E. 1992. Life. Sci. 51:1013-1019.
245. Marsault R., Vigne P., Brettmayer J.P., and Frelin C. 1999. J. Neurochem. 54:2142-2144.
246. Harvey A.L. 2001. Gen. Pharmcol. 28:712.
247. Harvey A.L. 2001. Brain Res. 762:223-227.
248. Dorandeu F., Wesherell J., Pernof-Marino J., Tattezsalle J.E., and Fosbraey P. 1997. J. Neurosci. Res. 48:499-506
249. Bagetta G., Nistico G., and Dolly J.O. 1992. Neurosci. Lett. 139:34-40.
250. Bondy C.A., and Russel J.T. 1988. Brain. Res. 453:397-400.

251. Awan K.A., and Dolly J.O. 1991. Neuroscience. 40:29-39.
252. Chilingarian A.M. 1960. Dokladi Acad. Sci. of Armenia. 31:241-243.
253. Hsu S.M., Roine L., and Farger, H.J. 1981. Histochem. Cytochem. 29:577-580.
254. Ambrisius X. 1987. Immunological methods. Ed.: Frimel. G.P. P. 14-15.
255. Zahir K.S., Zahir F.S., Thomas J.G., Dubrick S.J. 1999. Conn. Med. 4(9); 535-538.
256. Gillen C., Jander S., Stoll G. 1998. J. Neurosci. Res., 5(14); 489-496.
257. Westphalen R.A., Dood P.R. 1998. J. Neurochem., 71(Suppl.):15.
258. Rubinstein I., Abassi Z., Coleman R., Milman F., Winaver J.,, Better O.S. 1998. J. Clin. Invest. 10(6):1325-1333.
259. Dalkara T., Ayata C., Moskowitz M.A. 1999. Pages 85-96, *in* Inflammatory cells and mediators in CNS diseases, Harwood Academic Publishers.
260. Kevorkian G.A., Marukhyan G.L., Arakelyan L.N., Guevorkian A.G., and Galoyan A.A. 2001. Neurochem. Res. 26(7)827-830.
261. Kevorkian G.A., Kanayan A.S., Hayrapetyan R.L., Guevorkian A.G., Manukhyan G.L., Avanesyan S.A., Manukyan L.A., Voskanian L.H., and Galoyan A.A. 2001. Pages 86-94, *in* Galoyan A. (ed.), Proceedings of the International Conference. "Biochemical and Molecular Biological Aspects of Immune System", Encyclopedia Armenica Publishing House, Yerevan.
262. Wishnevski H.M., Ygbal K., and Dermott J.R. 1991. Neurotoxicology (park Forest II). 1:121-124.
263. Avtsyn A.P., Zhavoronkova A.A., Rish M.A., and Strockova L.S. 1991. Microelementosis. Medicine. Moscow.
264. Avtsyn A.P. 1996. Arch. Pathol. 48:5.
265. Avtsyn A.P. 1987. Clin. Med. 6:736.
266. Neil D., Leake A., Huges D. et. al. 1996. Neurochem. Res. 46:395-403.
267. Crapper-McLachlen D.R., and De Boni U. 1981. Neurotoxicology (Park Forest II), pp. 13-16.
268. Novak P., and Brus R. 1991. Postepy Hig. Med. Dosw. 50:621-633.
269. Alfey A.C., Le Gendre G.R., and Kaehny W.D. 1976. New England J. Med. 294:184-188.
270. Hodsman, A.B., Sherrard, D.J., Alfey A.C. et al. 1982. J. Clin. Endocr. 54:539-545.
271. Schakhlamov V.A., Polyakova G.P., Kondakova L.J., Chudinovskaya N.V., and Galoyan A.A. 2000. Morphologia 117:136.
272. Schakhlamov V.A., Galoyan A.A., Polyakova G.N., Vagradyan A.G., Simonyan M.A., Aghadjanov M.J., Bogdanova J.M., Altukova A.J., Kondakova L. 2002. Reports of Arm. Natl. Acad. Sci. 102(2):166-172.
273. Galoyan A.A., Kirakosova A.S., Saribekian G.A., Marukian T.Kh. 1977. Dokl. Acad. Nauk ArmSSR. 64(4):242-245.
274. Mandjikian S.P., Kirakosova A.S., Galoyan A.A., 1977. Dokl. Acad. Nauk ArmSSR. 64(5):291-294.
275. Mandjikian S.P., Kirakosova A.S., Karapetian R.O., Galoyan A.A., 1979. Dokl. Acad. Nauk ArmSSR. 69(2):115-118.
276. Galoyan A.A., Karapetian R.O. 1975. Dokl. Acad. Nauk ArmSSR. 60(4):232-236.
277. Guillemin R. 1975. Pages 117-121, *in*: Buniatian Ch. (ed.), Problems of brain biochemistry, Academic Press, Armenia, Yerevan Vol. 10:117.
278. Zavalishin I.A., Roichel V.M., Kondakova L.I. et al. 1998. J. Nevrolologii i Psychiatrii (in Rus.) 98(5):20-25.
279. Alov I.A., Aspiz M.E., Kazanceva I.A.1973. Moskow, Page 14.
280. Levi-Montalcini R., Angeletti P. 1968. Physiol. Evol. 48:534-569.
281. Barde Y.-A., Edgar D., Thoenen H. 1982. EMBO 1:549-553.
282. Leibrock J., Lottspeich F., Hohn A. et al. 1989. Nature 341:149-152.
283. Ernfors P., Ibanez C., Ebendal T. et al. 1990. Proc. Natl. Acad. Sci. USA. 87:5454-5458.
284. Hohn A., Leibrock J., Bailey K. et al. 1990. Nature. 344:339-341.
285. Berkemeier L., Winslow D., Kaplan D. et al. 1991. Neuron. 7:857-866.
286. Ip N., Ibanez C., Nye S. et al. 1992. Proc. Natl. Acad. Sci. USA. 89:3060-3064
287. Spenger C., Hyman C., Studer L. et al. 1995. Experimental Neurol. 133:50-63.
288. Hefti F., Hartikka J., Knusel B. 1989. Neurobiol. Aging. 10:515-533.
289. Lindsay R. 1994. Neurobiol. Aging. 15:249-251.
290. Seiger A., Nordberg A., Von Holst H. et al. 1993. Behav. Brain Res. 57:255-261.
291. Barres B.A., Burne J.F., Holtman B. et al. 1996. Mol. Cell. Neurosci. 9:146-156.
292. Frei E., Klusman I., Schnell L., Schwab M.E. 2000. Exp. Neurology. 163:373-380.
293. Liuzzi F., Lasek R. 1987. Science. 237:642-645.
294. Louis A.J., Magal E., Takajama S., Varon S. 1993. Science. 259:689-692.
295. McKeon R., Silver J., Large T. 1997. Exp. Neurol. 148:558-567.
296. Oderrfeld-Nowak B., Bacia A., Gradkowska M. et al. 1992. Neurochem. Int. 21:455-461.

297. Chekhonin V., Gurina O., Ryabukin I., Savchenko E., and Galoyan A.A. 2001. In Proceeding of the International conference "Biochemical and Molecular-Biological aspects of the brain Immune system (ed. A. Galoyan), pp. 150-145. Yerevan Encyclopedia Armenica Publisher House.
298. Marks N., and Berg M.J. 2001. Pages 146-158, *in* Galoyan A. (ed.), Proceedings of the International Conference. "Biochemical and Molecular Biological Aspects of Immune System", Encyclopedia Armenica Publishing House, Yerevan.
299. Cofman S.L., Halfer W., and Cole G.S. 2000. Mol. Cell. Neurosci. 15:183-198.
300. Medina V., Edmonds B., Young G.P. et al. 1997. Cancer Res. 57:3697-3707.
301. Litvak D.A, Evers B.M., Hwang K.O. et al. 1998. Surgery. 124:161-169.
302. Giuliano M., Lauricella M., Calvaruso G. et al. 1999. Cancer Res. 59:5586-5595.
303. Fischer I., Shea T.S., Sapirstein V.S. 1986. Neurochem. Res. 11:589-598.
304. Fischer I., Shea T.B., Sapirstein V.S., Kosik K.S. 1986. Brain Res. 390:99-109.
305. Koizumi H., Ohkawa I., Tsukahara T. et al. 1999. J. Pathol. 189:410-415.
306. Marks N., Berg M.J. 1999. Neurochem. Int. 35:195-220.
307. Marks N., Berg M.J., Saito M., Guidotti A. 1998. J. Neurosci. Res. 52:334-341.
308. Oberto A., Marks N., Evans H.L., Guidotti A. 1996. J. Pharmacol. Exp. Therap. 279:435-442.
309. Green D.R., Reed J.C. 1998. Science 281:1309-1312.
310. Stefanis L., Troy C.M., Qi H. et al. 1998 // J. Neurosci. V. 18. P. 9204-9215.
311. Lazebnik Y.A., Kaufmann S.H., Desnoyers S. et al. 1994. Nature. 371:346-347.
312. Orth K., Chinnaiyan A.M., Gard M. et al. 1996. J. Biol. Chem. 271:16443-16446.
313. Enari M., Sakahira H., Yokoyama H. et al. 1998. Nature. 391:43-50.
314. Lankin Y.Z., Gurevich S.M., and Burlakova E.B. 1975. In: Bioantioxidants. Moscow, pp. 73-78.
315. Stocks J., Dormandy T.L. 1971. Brit. J. Haemat. 20:95-111.
316. Grassl M., Maellering H. 1969. A. Z. Anal. Chem. Bd. 243(Supp.):416-423.
317. Kennedy E.P. 1962. Glycerokinase. Methods in Enzymology 5:476-481.
318. Beisenherz G., Bucher Th., Heinz K. 1955. Methods in Enzymology 1:391.
319. Nigh Y.M., Sanchez E.R. 1995. J. Steroid Biochem., and Mol. Biol. 52:187-194.
320. Veltishchev Yu.Ye., Yureva E.A., et al. 1981. Problems of Medical Chemistry. 4:441-449.
321. Brauns P., Velthuis H., Jansen P.G., Eijsman L., Hack C.E. 1998. J. Thorac Cardiovasc Surg. 115(4):949-951.
322. Twin M.M. et al. 2000. Biochemistry 39:3604-3611.
323. Tosh D., and Slack J.M.W. 2002. Nature Rev. (Molecular Biology). 3:187-192.
324. Biomson C.R.R., Rictze R.L., Reynolds B.A., Magli M.C., and Vescovi A.L. 1999. Science 283:539-5337.
325. Clarke D. et al. 2000. Science 288:1660-1663.
326. Orlic D. et al. 2001. Nature 410:701-705.
327. Mezey E., Chandross K.J., Harta G., Mald R.A., and McKercher S.R. 2000. Science 290:1779-1782.
328. Reyes, M. et al. 2001. Blood 98:2615-2635.
329. Priller S. et al. 2001. J. Cell. Biol. 155:733-738.
330. Priller J., Flugel A., Wehner T., Boentert M., Haas C.A., Prinz M., Fernandes-Kiett F., Prass K., Bechmann I., de Boer B.A., Frotscher M., Kreutzberg G.W., Persons D.A., and Dringal U. 2001. Nature Med. 7:1356-1361.
331. Gage F.H. 2000. Science. 287:1433-1438.
332. Eglitis M.A., and Mezey F. 1997. Proc. Natl. Acad. Sci. USA. 94:4080-4085.
333. Kondo T., and Raff M. 2000. Science. 289:1754-1757.
334. Neuman H., Wekerley H. 1998. J. Neuropathol. Exp. Neurol. 57:1-9.
335. Pawson T., Raina M., Nash P. 2002. FEBS Letters. 513:2-10.
336. Heldin C.-H., Osfman A., and Ronnstrand L. 1998. Biochem. Biophys. Acta. 1378:F79-F113.
337. Schlessinger J. 2000. Cell 103:211-225.
338. Ren R., Mager B.J., Cichetti P., and Baltimore D. 1993. Science 259:1157-1161.
339. Yu H., Chen J.K., Feng S., Dalgarno D.C., Brauer A.W., and Schrekiber S.L. 1994. Cell 76:933-945.
340. Sudol M., and Hunter T. 2000. Cell 103:1001-1004.
341. Koch C.A., Anderson D., Moran M.F., Ellis C., Pawson T. 1991. Science 252:668-674.
342. Sudol M., Sliwa K., and Russo, T. 2001. FEBS Letters 490:190-195.
343. Macias M.J. Wiesner S., Sudol M. 2002. FEBS Letters. 513:30-37.
344. Cesareni G., Ranni S., Nardelli G., Gastagnoil L. 2002. FEBS Letters 511:38-44.
345. Pawson, T., and Scott, J.D. 1997. Science. 278:2075-2080.
346. Feller, S.M. 2001. Oncogene. 20:6348-6371.
347. Li W., Fan J., and Woodley D.T. 2001. Oncogene. 20:6403-6407.

348. Reth M., and Wienands J. 1999. Gold Spring Harbor Symposia on Quantitative Biology. LXIV:326-328.
349. Moriggl R., Marine J.C., Topham D.J., Teglund S., Sexl V., McKay C., Piekorz R., Wang D., Parganas E. Yoshimura A., and Ihle J.N. 1999. Cold Spring Harbor Symposia on Quantitative Biology. LXIV:389-395.
350. Endo T.A. et al. 1998. Proc. Natl. Acad. Sci. USA. 95:144.
351. Hilton D.J., Richardson R.T., Alexander W.S., Viney E.M., Willson T.A., Sprigg N.S., Starr R., Nicolson S.E., Metcalf D. and Nicola N.A. 1998. Proc. Natl. Acad. Sci. USA, 95:144.
352. Starr R. et al. 1997. Nature 387:921.
353. Clemements J.L. et al. 1998. Science. 281: 416.
354. Jackman J.K., Motto D.G., Sun Q., Tanemoto M., Turck C.W., Peltz G.A., Koretzky G.A., Findell P.R. 1995. J. Biol. Chem. 270(13):7029-7032.
355. Wienands J., Schweikert J., Wollscheid B., Jumaa H., Nielsen P.J., Reth M. 1998. J. Exp. Med. 188(4):791-795.
356. Shi J., Ross Ch.R., Leto T., and Blecha F. 1996. Proc. Natl. Acad. Sci. USA. 93:6014-6018.
357. Finco T.S., Yablonski D., Lin J., and Weiss A. 1999. Gold Spring Harbor Symposia on Quantitative Biology. LXIV:265-274.
358. Morgan B., Thorpe L.W., Marchetti D. and Perez-Polo J.R. 1989. J. Neurosci. Res. 23:41-45.
359. Whittemore S.R., Seiger A. 1987. Brain Res. Rev. 12:439-464
360. Thorpe L.W., Morgan B., Beck C., Werbach-Perez K., Perez-Polo J.R.. 1988. Pages 558-594, *in:* Cellular and Molecular aspects of neuronal development and regeneration. Heidelberg: Springer-Verlag.
361. Matsuda H., Coughlin M.D., Bienstock J., Denburg J. 1988. Proc. Natl. Acad. Sci. USA. 85:6508-6512.
362. Thorpe L.W., Werbach-Perez K., Perez-Polo J.R. 1987. Ann. N.Y. Acad. Sci. 496:310-311.
363. Gee A.P., Boyle M.D.P., Munger K.L., Lawman M.J.P., Yung M. 1983. Proc. Natl. Acad. Sci. USA. 80:7215-7218.
364. Perez-Polo J.R., Werbach-Perez K. 1988. Pages 394-410, *in*: Neuronal development and regeneration cellular and molecular aspects, series II, Heidelberg, Springer-Verlag, Vol. 22
365. Andress R.Y., Jeng I., Bradshaw R.A. 1977 Proc. Natl. Acad. Sci. USA. 74:2785.
366. Serri I., Rogalski P., et all. 2000, J. Biol. Chem. 275(33):25082-25088.
367. Gurvits B.Ya., Tretyakov O.Yu., Klishina, N.V., Stoeva S., Voelter, W., and Galoyan A.A. (2000). Neurochem. Res. 25 (8), 1125-1129.

ABBREVIATIONS

Ab - antibody
ABC - avidin-biotin peroxidase complex
ACE - angiotensin-converting enzyme
ACP - antivenin (crotalidae) polyvalent
ACTH – adrenocortinotrophic hormone
AD – Alzheimer's disease
ADCC - antibody-dependent cell cytotoxicity
AFC - antibody-forming cells
AML – acute myeloid leukemia
AML 1 (transcription factor)
APP - amyloid precursor protein
ATCC – American Type Cell Collection
ATP – adenosine-triphosphate
BA - background activity
BBB - blood-brain barriers
BDNF - brain derived neurotrophic factor
BSA – bovine serum albumin
C – cytosol
C-99 - 99-residue polypeptide CTF
cAMP – cyclic adenosin monophosphate
CC - cumulative curves
CD – cluster of differentiation
CFC - colony-forming cells
CFU – colony-forming unit
cGMP – cyclic guanosin monophosphate
ChS – chondroitin sulfate
c-Jun - proto-oncogene.
CL – cardiolipin
CLP - common leukocyte progenitor
CMP - common myeloid progenitors
ConA - concanavalin A
α-GP - glycerophosphates
CPA – cyclophosphamide
cpm – counts per minute
Cpu – capsula interna
CRF - corticotropin releasing factor
Crk – symbol for, a protooncogene related to t-crk, the oncogen of avian sarcoma viruses CT10.
CS - crush syndrome
CSF - cerebrospinal fluid
CTF - cell-associated C-terminal fragment
CTF CAAT box-binding transcription factor or nuclear factor 1 (NF-1); any of a class of transcription factors
CTLL-2 - IL-2 dependent line of the cytotoxic T-lymphocytes
CV - cobra venom
DAG - diacylglycerides
DEX - dexamethasone
DMEM – Dulbecko's modified eagle's medium
DMN - dorsal motor nucleus of vagus
DN - double negative
DOAP - dihydroxyacetone phosphate
DP - double positive
DPG - diphosphatidilglycerides
ELISA – enzyme-linked immunosorbent assay
EMV - encephalomyocarditis virus
Epo – erythropoietin
ER - endoplasmic reticulum
Erg-1 - gene
ERK 1- extracellular signal regulated kinase
Fab-molecules (CroFab, FabAV and ACT)
Fas-L - member of the TNF receptor family
FBS - fetal bovine serum
FCS - fetal calf serum
FITC – fluorescein isothyocyanate
FP - field potential
FTOC - fetal thymus organ culture
Fyn – a gene family encoding nonreceptor tyrosine kinases of the Src family that are implicated in the control of cell growth
GABA - gamma-amino-n-butyric acid
GAG – glycosaminoglycan
GATA – transcriptional activator protein that serves as a general switch factor for crythroid development
G-CSF – granulocyte-colony-stimulating factor
G-CSFR - granulocyte CSF receptor
GDNF – glial-derived neurotrophic factor
GFAP - glial fibrillary acidic protein
GH - growth hormone
GH-RH – growth hormone-releasing hormone
GK - glycerokinase
GM-CSF – granulocyte-macrophage-colony stimulating factor
G-MCSFR - granulocyte-macrophage-CSF receptor
GPD – glycerophosphate dehydrogenase
Grb2 growth factor receptor-bound protein 2; also called Ash protein
GRP - gastrin-releasing peptide
GTP - guanosine triphosphate
HBSS - Hanks balanced salt solution
HII - histogram of interimpulse intervals
HPA axis – hypothalamus pituitary-adrenal axis
HPLC - high performance liquid chromatography
HS - heparin sulfate proteoglycans

HSC – hematopoietic stem cell
i/m – intramascular
i/p - intraperitonealy
i/v – intravenously
IFN - interferon
IKAROS – a protein that binds and activates the enhancer of a gene CD3-δ involved in T-lymphocyte specification and maturation
IL – interleukin
IN - interneuron
IphF – immunophilin fragment
IU - international units
Jak - Janus protein kinase
LAK - lymphokin-activated killers
LAT – linker for activation of T-cells
LD – lethal dose
Lex A a gene in escherichia coli encoding a repressor that regulates the genes induced by SOS repair
LH-RH – luteinizing hormone-releasing hormone
LPC – lysophosphatidylcholine
LPO – lipid peroxidation
LPS – lipopolysaccharide
LTP – long-term potentiation
Lyn – a protooncogene encoding a tyrosine kinase belonging to the Src family
mAb - monoclonal antibody
M-CSF – macrophage-colony stimulating factor
M-CSFR - macrophage colony-stimulating factor receptor
MF – macrophage
MHC - major histocompatibility complex
MIF - macrophage migration inhibitory factor
MLCK – myosin light chain kinase
MN - motoneuron
MNC - mononuclears
MP1 – monophosphoinositides
myb – a gene family that ancodes DNA-binding proteins, contains DNA-binding, transcription activation.
MZF 1 - myeloid zinc finger domains
NADPH – nicotinamide-adenine dinucleotide phosphate (reduced form)
NAS – normal goat serum
NF-kB – a transcription factor for eucaryotic RNA polymerase II promoter
NGF – nerve growth factor
NGH - neurosecretory granules of hypothalamus
NGN - neurosecretory granules of neurohypophysis
NGS - normal goat serum
NMDA – N-methyl-D-aspartate
NO - nitric oxide
NPV – Nucleus Paraventricularis
NPY - neuropeptide Y
NSO – Nucleus Supraopticus
NT-3 – neurotrophin-3
NT-4/5 - neurotrophin 4/5
NTOC - neonatal thymus organ culture
NTS – nucleus tractus solitarius
NVAG - neurophysin-vasopressin-associated glycoprotein
OAF - oxygen active forms
P1 – phosphatidilinosites
P1P2 – phosphatidylinositol 4,5-bisphosphate
PA – phosphatidil acid
PAP – peroxidase-antiperoxidase
Par. 1 – pariental field
Pax – paired box genes encode a family of transcriptional regulators
PBS – phosphate-buffered saline
PC – phosphatidylcholine
PDE – phosphodiesterase
PE – phosphatidylethanolamine
PETH – Peri-Event Time Histogram
PHA – phytohaemagglutinin
PHC - protein-hormonal complexes
PI – phosphatidylinosites
PIP - phospholipase inhibitor of Python
PL – phospholipids
PLCγ-1 – phospholipase Cγ-1
PMF – peritoneal macrophage
PMSF – phenylmethylsulphonyl fluoride
PR - proline-arginine
PRP – proline-rich_peptide
PRP-1-IR - PRP-1-immunoreactivity
PS – phosphatidylserine
PTK – protein tyrosine kinase
PU.1 – transcription factor (a major regulator of myeloid development is a member of the Ets transcription family)
RTK - receptor tyrosine kinases
s/c - subcutaneous
SC - spinal cord
SCF – stem-cell factor
SH2 domain – Src homlogy 2 domain
SLP-76 - leukocyte-specific-adaptor protein
SM – sphingomyelin
SP - single positive
SPC3 - subtilisin-like prohormone convertase
Src-2 – (an oncogene (v-src), originally discovered in the Rous sarcoma, virus protooncogene codes for a related protein's
STAT – signal transducer and activator of transciption
TACE – TNF-converting enzyme
TCD - tissue cytopathogenic doses
TCR – T-cell receptor

αβTCR - αβ T cells receptor
TCR/CD3 - T-cell-antigen receptor complex
TF – transferrin
TFA - trifluoroacetic acid
TH1 - T-helper type 1 cells
TH2 - T-helper type 2 cells
Thr – threonine
TNF - tumor necrosis factor
Tpo – trombopoietin
TRH - thyrotropin-releasing hormone (thyreoliberin)
Trk – a protooncogene the product is a tyrosine kinase receptor
TrkA, TrkB and TrkC – members of Trk family of tyrosine kinases
tRNA – transport ribonucleic acid
TSLP - thymic stroma derived lymphopoietin
Tyk – tyrosine kinase
Tyr – tyrosine
VIP - vasoactive intestinal polypeptide
VR - Vipera Raddei
WW domain – proline-rich peptide binding domain
YXXP – Tyr-XX Pro
ZAP-70 – tyrosine kinase of Syk family (a 70 kD protein-tyrosine kinase that associates with TCRξ chain)

ACKNOWLEDGEMENTS

I should like to express my sincere gratitude to Prof. Abel Lajtha for long years of collaboration and valuable advice given in the process of preparation this book for publication.

Cordially thanks to my coworkers and other collegues taking part in our investigations.

I should also like to express my gratitude to Prof. Alexander Preobrazhenski for the help of the preparation of this manuscript.

I am thankful to K. Gevorgyan, K. Galoyan, and R. Sulkhanyan for their help in international communications.

INDEX

GPSR Compliance

The European Union's (EU) General Product Safety Regulation (GPSR) is a set of rules that requires consumer products to be safe and our obligations to ensure this.

If you have any concerns about our products, you can contact us on ProductSafety@springernature.com

In case Publisher is established outside the EU, the EU authorized representative is:

Springer Nature Customer Service Center GmbH
Europaplatz 3
69115 Heidelberg, Germany

Batch number: 10607924

Printed by Printforce, the Netherlands